Heba Salah Hamed
Adel Mahdy Shaheen
Mariam Naguib Okal

Parsley nanoparticles against zinc oxide nanoparticles in Nile tilapia

Heba Salah Hamed
Adel Mahdy Shaheen
Mariam Naguib Okal

Parsley nanoparticles against zinc oxide nanoparticles in Nile tilapia

Noor Publishing

Imprint

Cover image: www.ingimage.com

Publisher:
Noor Publishing
is a trademark of
Dodo Books Indian Ocean Ltd. and OmniScriptum S.R.L publishing group

120 High Road, East Finchley, London, N2 9ED, United Kingdom
Str. Armeneasca 28/1, office 1, Chisinau MD-2012, Republic of Moldova, Europe
Managing Directors: Ieva Konstantinova, Victoria Ursu
info@omniscriptum.com

Printed at: see last page
ISBN: 978-620-9-28700-8

Courses

Courses studied by the candidate in partial fulfillment of the requirements for the degree of M.Sc.

1-Physiology

2- Histology

3- Immunology

4- Ethics of scientific research

5- Writing scientific research

Acknowledgement

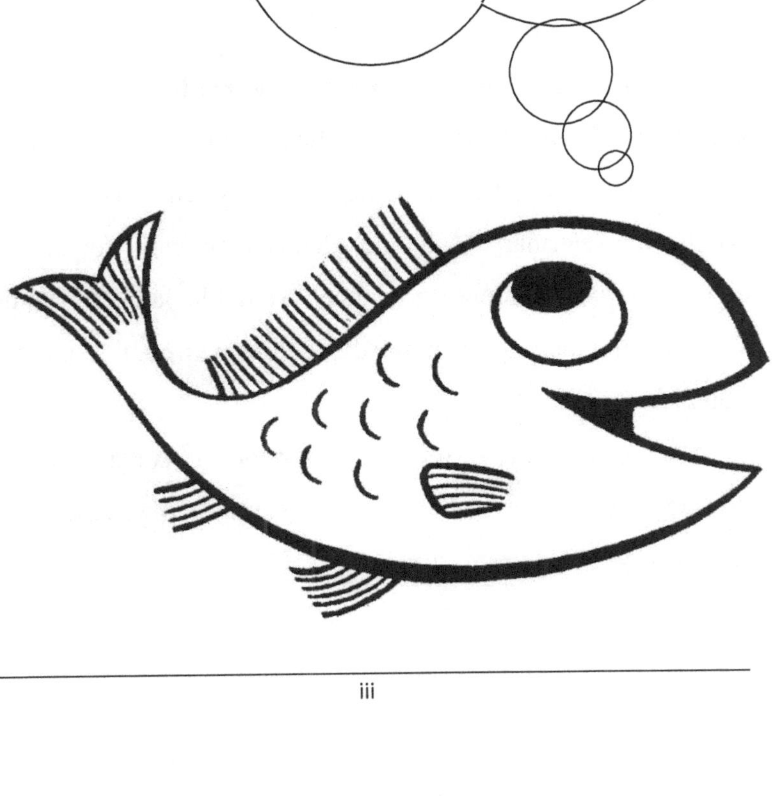

ACKNOWLEDGMENT

Praise be to Allah who extends in knowledge whom be wills I am extremely grateful to Allah who gave me the ability to finish this work.

I wish to express my deepest gratitude to **Prof. Dr. Heba Salah Mohamed Hamed,** Professor of Fish Physiology, Zoology Department, Faculty of Women for Arts, Science & Education Ain Shams University, for her careful guidance, supervision, kind encouragement, continuous support, and instructive criticism, she suggested the subject, set up the plan and offered brilliant ideas from the start to the end.

My thanks and sincere appreciation to **Prof. Dr. Adel Abdel-Aleem Shaheen**, Professor of fish animals ,Diseases and Management, Aquatic Animals Diseases and Management, Faculty of Veterinary Medicine, Benha University for his keen supervision, continuous guidance, valuable advice, endless donation, and facilitating all the difficulties while supervising this work.

My cordial thank for all members of Aquatic Animal Medicine Department, Faculty of Veterinary Medicine, Benha University,

My deepest thanks to all members of the Zoology Department, Faculty of Women for Arts, Science & Education, Ain Shams University, for their help and cooperation.

Dedication

I dedication this work to my family ❤

Mariam

Abstract

ABSTRACT

Effect dietary parsley leaves extract and parsley nanoparticles against zinc oxide nanoparticles toxicity in Nile tilapia (*Oreochromis niloticus)*

Zinc oxide nanoparticles (ZnO-NPs) are effectively used in varied aquaculture applications such as Electronics, cosmetics, catalytic, ceramic, pigments, glass, rubber, Baby powders, antidandruff shampoos, and fabric treatments. This study aimed to determine LC_{50} of (ZnO-NPs) in Nile tilapia, (*O.niloticus*), as well as the protective role of parsley leaves and parsley nanoparticles against ZnO-NPs toxicity.

Fish were divided into 6 groups of triplicates as follows:

group I: was kept as a control,

group II: was fed basal diet and parsley leaves extract (4 g/kg diet) for 6 weeks

group III: was fed a free basal diet and parsley nanoparticles (4 g/kg diet) for 6 weeks

group IV: exposed to 8 mg/L ZnO-NPs and fed with a basal diet for 6 weeks

group V: exposed to 8 mg/L ZnONPs and fed with parsley leaves extract (4 g/kg diet) for 6 weeks.

group VI: exposed to 8 mg/L ZnO-NPs and fed with parsley nanoparticle extract (4 g/kg diet) for 6 weeks.

At the end of the experiment, blood was collected for biochemical analysis. As well as liver tissues of Nile tilapia were taken for oxidative enzymes and histopathological studies.

The results exhibited that exposure of *O. niloticus* to 8 mg/L ZnO-NPs lead to a marked increasing in liver enzymes Alkaline phosphatase (ALP), alanine aminotransferase (ALT), and Aspartate aminotransferase (AST). Also, these increases were indicated in kidney activities (urea and creatinine)

Levels of total proteins, albumin, and globulin were markedly significantly increased. However, the A/G ratio decreased. Also, glucose, cortisol, total lipids, and cholesterol levels were increased and triglycerides were significantly decreased.

These results indicated that the levels of reduced glutathione (GSH), catalase (CAT), superoxide dismutase (SOD), Nitric oxide (NO), and Glutathione Peroxidase (GPx) were significantly decreased, while total antioxidant capacity (TAC) and Malondialdehyde (MDA) increased after exposure to 8 mg/L ZnO-NPs for 6weeks, compared to the control fish

Analysis of fish blood samples indicated a significant decrease in WBCs, monocytes, MCV, MCH, and platelets. On the other hand, an increase in hemoglobin content and MCHC were increased significantly.

Histological study in hepatic tissues of Nile tilapia represented more damage and alternation in tissues that were exposed to 8 mg/L ZnO-NPs for 6 weeks compared to the control group and treatment group with a parsley leaves and parsley nanoparticles (4g/kg diet).

From this study, we investigated that co-administration of Nile tilapia on (4g/kg diet) parsley leaves or parsley nanoparticles enhanced the destructive effects of ZnO-NPs. It was more evidence in fish co-administered with parsley nanoparticles.

Key words: Zinc oxide nanoparticles (ZnO-NPs), Nile tilapia, parsley *(Petroselinum crispum),* parsley nanoparticles, Biochemistry, Histopathology.

Content

CONTENT

LIST OF TABLES

LIST OF FIGURES

LIST OF ABBREVIATIONS

ALT	Alanine amino transferase
ALP	Alkaline phosphatase
AST	Aspartate amino transferase
BUN	Blood urea nitrogen
°C	Degrees celsius
C2C12	an immortalized mouse myoblast cell line
CAT	Catalase enzyme
Cells/mL	cells per microliter
EDTA	Ethylene diamine tetraacetic acid
GSH	Reduced glutathione
H	Hour
Hb	Hemoglobin Concentration
HDL	high-density lipoprotein
G	Gram
g/l	Gram per liter
g/dl	Gram per deciliter
GPx	Glutathione Peroxidase
H&E	Hematoxylin and Eosin
IgG	Immunoglobulin G
IgM	Immunoglobulin M
IP	Intraperitoneal
IV	Intravenous
Kg	Kilogram
LC_{50}	Half lethal concentration
LDL	low-density lipoproteins
LPO	Lipid peroxidation
MCH	mean corpuscular hemoglobin
MCHC	mean corpuscular hemoglobin concentration
MCV	mean corpuscular volume
MDA	Malondialdehyde
mg/dl	Milligram per deciliter

mg/kg	Milligram per kilogram
mg/L	Milligram per liter
µ/L	Micro per liter
µg/L	Microgram per liter
Mm	Micrometer
µg/dl	Microgram per deciliter
NO	Nitric oxide
pg/cell	Picograms per cell
RBCs	Red blood cells
ROS	reactive oxygen species
SGR	Specific growth rate
SiO_2	Silicon dioxide
SOD	Superoxide dismutase
TAC	Total Anti-oxidant Capacity
TEM	transmission electron microscopy
TiO_2	Titanium dioxide
UV	Ultraviolet
VLDL	very-low-density lipoprotein
W	Weight
WBCs	White blood cells
ZnO	Zinc oxide
ZnONPs	Zinc oxide nanopartical

Introduction

INTRODUCTION

Recently, nanotechnology has emerged as an excellent field of technology **(Kalnie *et al.,* 2006),** and increasing the demand for nanoparticles in many applications such as industrial applications, biological and biomedical science, health care technology, and household appliance **(Benn and Westerhoff, 2008; Masciangioli and Zhang, 2003; Nohynek *et al.*, 2007; Rajkumar *et al.*, 2016).** Although the application of nanoparticles supports our life at the same time, its impact on the environment and living organisms are becoming an important issue **(Adhikari *et al.*, 2012; Yousefian and Payam, 2012).**

Since the last decade, the field of manufacturing zinc oxide particles (ZnO-NPs) have been increased **(Kaya *et al.*, 2015)**. Production of ZnO-NPs has the highest third position after Titanium dioxide (TiO2) and Silicon dioxide nanoparticles (SiO2-NPs) **(Piccinno *et al.*, 2012)**. ZnO-NPs are widely applied in electronics, cosmetics, catalytic, ceramic, pigments, glass, and rubber due to their UV absorbance properties **(Wang *et al.*, 2004, Ma *et al* 2013; Zhao *et al.*, 2013)** and are also used as baby powders, antidandruff shampoos and fabric treatments for UV shielding **(Bechrii *et al.*, 2008)**. Since ZnO-NPs are widely used, high concentrations of the metal ions released to the aquatic environment and their bioaccumulation could be toxic to the aquatic fauna, particularly, fishes which in turn hamper fish farming **(Abdel-Daim *et al.*, 2018)**. ZnO-NPs induced adverse effects in zebrafish and fish larva, hatching rate, and malformation **(Zhu *et al.*, 2008)**.

Fish are widely used as indicators for environmental pollution and to evaluate the health of the aquatic ecosystem and physiological changes. Therefore, **Rajkumar *et al.*, (2016)**. Therefore, it is used as a biological indicator of ecotoxicological studies.

In teleosts, reactive oxygen species (ROS) are generated by external environmental factors which could lead to oxidative stress and free radicals) **(Li *et al.*, 2008)**, increasing lipid peroxidation (LPO). Furthermore, ZnO-NPs, like other nanoparticles can cause toxicity by increasing the concentration of intracellular ROS and decrease in antioxidant levels **(Hussain *et al.*, 2008; Singh *et al.*, 2008; Rajkumar *et al.*, 2016)**.

Several investigators documented that ZnO-NPs could induce oxidative stress in the fish body by disturbing the prooxidant/antioxidant balance of cells **(Kaya *et al.*, 2015; Abdelazim *et al.*, 2018; Abdel – Daim *et al.*, 2018)**.

Parsley belongs to the Umbelliferae family. Parsley originated in the Mediterranean; however, it is currently cultivated worldwide **(Kolarovic *et al.*, 2010)**. Phenolic compounds, particularly flavonoids (e.g,. apigenin, apiin, and 600 -acetylation), coumarins, furocoumarins, and essential oil components mainly (myristicin and apiol) are the active components that have been isolated from parsley Moreover, ascorbic acid, tocopherols, and carotenoids have also been isolated **(Farzaei *et al.*, 2013)**

Parsley has many pharmacological effects, including antioxidant, hepatoprotective, and cardioprotective effects **(Popovic *et al.*, 2007, Sener *et al.*, 2003, Abdellatief *et al.*, 2017)**. In addition to nephroprotective, neuroprotective ,anti-diabetic, analgesic, spasmolytic, laxative, and diuretic effects, and antibacterial and antifungal activities **(Farzaei *et al.*, 2013)**.

Nile tilapia, *Oreochromis niloticus* has an important place in the world of fish farming. Tilapia fish has high tolerance to stress and diseases, and hence, is used as a model organism in toxicological studies **(Kaya *et al.*, 2013&2015)**.

Studies show that zinc oxide can activate immune cells, whether this exposure is harmful effects on systemic immunity are uncertain, and numerous studies have begun to be interested in the effect of zinc oxide on immune cells at both sub-toxic and toxic levels. **(Andersson-Willman *et al.*, 2012; James *et al.*, 2013; LozanoFernandez *et al.*, 2014; Luo *et al.*, 2014; Shen *et al.*, 2013).**

Oxidative stress and toxicity of cells as well as in blood cells in vivo and the laboratory may be due to the structure and size of nanoparticles **(Koohi *et al.*,2017; Xia *et al.*,2006; Ruiz *et-al.*,2014)** Nanoparticles induce hemolysis, activated coagulation, granulocytes, induced inflammation and hemolysis in human blood samples. **(Chang *et al.*, 2012; Mayer *et-al.*,2008).**

AIM OF THE WORK:

The study was designed to investigate the toxic effects of Zinc oxide nanoparticles (ZnO-NPs) on ***Oreochromis niloticus,*** *regarding:*

1-Estimation of the half-lethal concentration of ZnO-NPs for 96h.

2-Investigation of clinical and behavioral variations.

3-Determination of Haematological and biochemical parameters and the oxidative stress

4-Determination of innate immune parameters.

5- Histopathological alterations of liver tissues.

6- The possible ameliorative effect of parsley leaves extract and parsley nanoparticles against ZnO-NPs toxicity.

Review
of
literature

REVIEW OF LITERATURE

Zinc oxide nanoparticles (ZnO-NPs):

ZnO-NPs and half lethal concentration (LC_{50}):

Lopes *et al.*, (2013) indicated that the 48-hour median lethal concentration (LC_{50}) was 1.32 mg/L for ZnO-NPs from 80 nm to 100 nm in *Daphnia magna*, also **Alkaladi *et al.*, (2015)** found that the results showed that 86 h LC_{50} was 3.1 ± 0.4 mg/L in *O. niloticus* exposed to ZnO -NPs.

According to **Taherian *et al.*, (2018)** detected the synthesized ZnO-NPs from Satureja hortensis on rainbow trout, *Oncorhynchus mykiss,* where three groups were exposed to different concentrations (1, 10, 100 mg/L) of ZnO-NPs. The value at 86 hours turned out to be 25.50 mg/L for trout, making these nanoparticles a low-toxicity material in terms of toxicity classification and green nanoparticles is much less risky. **Alkaladi *et al.*, (2020)** used ZnO-NPs concentrations ranging from zero to 8.5µg/ L, the resulting mortality LC_{50} for Zno-NPs in *O. niloticus* calculated after 96 h. In the same year, **Dobretsov *et al.*, (2020)** indicated that the highest toxicity was observed for ZnO-NPs (LC_{50} ~15 mg/L) using the microalgae *Dunaliella salina*, the salt shrimp, *Artemia salina*, and the marine bacterium, *Bacillus cereus.*

ZnO-NPs and clinical signs and behavioral effects:

Kavitha and Rao (2007) observed that in mosquito fish (*Gambusia affinis*) exposed to (20.48 mg/L.) monocrotophos for 86h, erratic swimming, loss of feeding, and loss of equilibrium. **Chan *et al.*, (2014)** found that a Similar hatching delay that did not cause larval mortality or malformation was observed when exposed to ZnO-NPs suspensions to fertilized zebrafish eggs (*Danio rerio*) were exposed to different concentrations of ZnO-NPs suspensions (control, 0.1, 0.5, 1, 5, and 10 mg/L) were measured until they reached the free-swimming stage. Their average and maximum speed were changed in high ZnO-NPs concentrations (5 and 10 mg/L).

Moreover, **Kaya *et al.*, (2015)** showed behavioral changes such as inertia, ptosis, and desire to hide have been observed when tilapia fish were exposed to different concentrations of ZnO-NPs (1 mg/L and 10 mg/L) in *O. niloticus*. Also, the study by **Hamed and Osman (2017)** noted the loss of equilibrium was followed by handing vertically in the water and irregular erratic swimming and convulsion, in African catfish (*Clarias gariepinus*) exposed to carbofuran for one month.

In the same fish, **Naguib *et al.*, (2020)** observed a change in the color of the skin, abnormal behavior such as fast swimming, and loss of equilibrium after exposed that 100 mg/L Ag-NPs for 15 days in *C. gariepinus*. Recently, **Khan *et al.*, (2022)** recorded abnormal movements, loss of body balance and body control, increased swimming speed, and side swimming, in that for 86 hours, when *Oreochromis mossambicus* exposed to 20 ppb ZnO-NPs

ZnO-NPs and liver function:

Lee *et al.*, (2014) showed that when common carp, *Cyprinus carpio*, were exposed to ZnO-NPs at a rate of 2.4 ml/L, there was a significant increase in AST levels compared to the control levels. And **Pandurangan and Kim (2015)** found that there was an increase of 82.3%, 115.4%, and 168.2% ALT at 48 h, while by 50%, 116.7%, 150% at 72 h when C_2C_{12} cells were exposed to different concentrations (2, 3 and 4 mg/ml) of ZnO-NPs, While AST increased by 45%, 70% and 125% at 48 hours and by 38.5%, 73% and 107.7% at 72 hours when C_2C_{12} cells were exposed to the same concentrations, Also, an increase in ALP by 37.1%, 68.6% and 100% in 48 hours, while by 45%, 72.5%, 81% at 72 hours in the same concentrations.

Ansar *et al.*, (2018) observed that AST, Alt increased when rats were exposed to zinc oxide with a diameter less than 100 nm by (600 mg/kg bw). Moreover, **Abdelazimetal *et al.*, (2018)** detected that when exposed to slow fish at 50 mg/liter, liver enzymes ALT and AST were increased in Nile tilapia exposed to 50 mg/L.

Recently, **Sudhakaran *et al.*, (2020)** showed that AST and ALT were significantly increased when mice were exposed to intravenous (IV) and intraperitoneal (IP) ZnO-NPs with nanoparticle size equal to 20 nm confirming strong hepatotoxicity.

ZnO-NPs and Kidney function:

Yan *et al.,* (2012) found that the Mice which treated with ZnO- NPs (100, 300, and 1000 mg/kg respectively),showed significant increases in serum creatinine and BUN were observed in the 1000 mg/kg group. other investigators

Alkaladi *et al.,* (2015) found that the ratio of Creatinine and urea increased exposure of *O. niloticus* to zinc oxide by (1 mg/l, 2 mg/l) for 7 days and the increase was greater after 15 days. In this concern, **Dehkourdi *et al.,* (2017)** showed that When rats were exposed to doses of ZnO-NPs (50, 100, 200) mg/kg, Blood Urea Nitrogen (BUN) and creatinine were increased, especially in the expensive concentration (200 mg/kg), which indicates its high toxicity for 30 days.

Yousef (2018) showed that A group of male albino rats aged between 4-5 months (100 mg/kg body weight) was given ZnO-NPs (particle size of 100 nm) for 75 consecutive days. The creatine and urea were signficatly increased. **Munir *et al.,* (2020)** found that when *Catla catla* fish treated with ZnO-NPs (0, 25, 50, 75, and 100 μg/g body weight) for 30 days,the urea concentration increased with a decrease in creatinine.

In the same year, **Rahdar *et al.,* (2020)** demonstrated that when mice were injected with doses of (5, 50, and 100 mg/kg) of ZnO-NPs, twice a week for 8 weeks, the creatinine, urea levels were increased, especially in the last two high concentrations.

ZnO-NPs and Total proteins and immune function:

According to **Wang *et al.*, (2010)** recorded that total protein and A/G ratio decrease in mice exposed to zinc oxide nanoparticles, at 2.5 mg/kg body weight by inhalation, twice daily for 3 days. Whenever, **Saman *et al.*, (2013)** showed no change in total protein level when mice exposed to 100, 200, and 400 mg/kg ZnO-NPs, respectively, 14 days.

Abdel-Khalek *et al.*, (2015) showed that when *O. niloticus* exposed to ZnO-NPs (0.08 g/L) for 28 days there was a decrease in the protein level. And, **Abdelazim *et al.*, (2018)** observed that, the liver enzyme's total protein and albumin of were *O. niloticus* exposed to 50 mg/L ZnO-NPs decreased significantly

Banaee *et al.*, (2018) showed no significant differences in plasma total protein, albumin, and globulin levels common carp, *C.carpio* exposed to (5,10, and 15 mg/kg) by food for 21 days compared to the control group.

ZnO-NPs and Stress indicators:

Reza *et al.*, (2014) stated that the injection of a group of rats with ZnO-NPs (5,10, 20, and 40 mg/kg) body weight, caused a decrease in cortisol level. But, **Abdel-Khalek *et al.*, (2015)** found that there was an increase in glucose level when *O. niloticus* exposed to Zn-NPs in a percentage (0.08 g/l) for 28 days, where the size of ZnO-NPs was (<50).

Farsania *et al.*, (2017) showed that when Nile tilapia was exposed to ZnO-NPs at a rate of 1.5 and 2.5 mg/L, an increase in glucose and cortisol levels was observed, where the experiment continued for 10 days. **Banaee *et al.*, (2018)** showed that 10 and 15 mg/ kg ZnO-NPs in the diet caused a marked increase in plasma glucose level of common carp for 21 days.

In the same line, **Mondal *et al.*, (2020)** reported that when *Labio rohita* exposed to ZnO-NPs at 10, 20, and 30 mg/kg respectively for 45 days. The results showed that fish fed on 20 mg ZnO-NPs per kg had the highest glucose level as the particle size of the synthesized ZnO-NPs was close to 100 nm. **Alkaladi *et al.*, (2020)** recorded that cortisol levels increased when Nile tilapia exposed to 1/20and 1/30 LC_{50} of ZnO-NPs (2.05 and 1.23µ/l) for four days.

ZnO-NPs and lipids parameters:

A study conducted by **Yousafzai and Shakoori (2011)** showed increased in cholesterol levels against aquatic pollution in freshwater fish. This disagrees with **Lee *et al.*, (2014)** who found that there was no significant change in the levels of total cholesterol and triglycerides when Young common carp exposed to at (0.1, 0.3, 0.8, and 2.4 mg/L) ZnO-NPs for 12 weeks.

Fahmy *et al.*, (2014) observed that total lipid and cholesterol levels were increased when the freshwater snail, *Biomphalaria alexandrina,* exposed to zinc oxide (7 and 35 µg/ml) for three weeks (< 35 nm average particle size). And**, Abdel-Daim *et al.*, (2018)** observed an increase in cholesterol levels of Nile tilapia fish exposed to (50 mg/L) of ZnO-NPs for 30 days

Kovacik *et al.*, (2018) stated that after exposure *C. carpi* to trace metals, a significant increase in cholesterol levels where observed this agrees with **Mahboub *et al.*, (2022)** who detected that *C. Carpio* was exposed to 13 mg/kg or 26 mg/kg of ZnO-

NPs for 30 days, showed marked increase in triglycerides and cholesterol levels.

ZnO-NPs and Oxidative stress and Anti-oxidants enzymes:

Catalase enzyme (CAT):

Lee *et al.*, (2012) observed increased CAT activity in the liver of common carp (*C. carpio*) when exposed to (200 mg/L) Ag-NPs. But results by **Wu and Zhou (2013)** showed that when *Oryzias latipes* exposed to silver nanoparticles (Ag-NPs) after 14 days, noted CAT levels were decrease activity.

Benavides *et al.*, (2016) observed that when freshwater fish (*Carassius auratus*) were exposed to (10,100 µg/L of ZnO-NPs) for 20 days, CAT increased on the 14th day. But **Abdelazim *et al.*, (2018)** found that Nile tilapia were exposed to 1 and 2 mg/L ZnO-NPs, and a reduction in (CAT) was observed, as muscle was collected on the 7th and 15th days of the experiment.

Shahzad *et al.*, (2018) showed that when tilapia *(O. mossambicus)* was exposed to 1.5 mg/L ZnO-NPs for 14 days, an increase in the CAT was observed in the gills, but decreased in the liver. Recently, **Aziz *et al.*, (2020)** Catalase activity decreased when exposed *Labeo rohita* to ZnO-NPs for 80 days with a percentage of 31.15 and 57.84 mg/L, and the zinc oxide nanoparticle was of size was 52.22 nm.

Glutathione content (GSH):

Jin *et al.*, (2010) noted that zebrafish (*D. rerio*) were exposed to (1, 10, 100, and 1000 µg/l) atrazine for 2 weeks, and GSH was decreased. According that **Xiong *et al.*, (2011)** demonstrated that when zebrafish were exposed to 5 mg/L ZnO-NPs for 7 days, an increase in the percentage of GSH in gut tissue was observed.

In (2013), Fernández *et al.*, showed that when exposure the gonads (RTG-2), liver (RTL-W1), and liver hepatoma (RTH-148) of rainbow trout (*Oncorhynchus mykiss*) to ZnO-NPs concentrations (0.78,1.56, 3.12, 6.25, 12.5 and 25 µg/mL), Increased GST activity was observed with increasing ZnO-NPs (RTG-2 of 6, 24 and 48 hours of exposure and RTH-148 and RTL-W1 for 6 and 24 hours of exposure). And **Kaya *et al.*, (2015)** found that (*O. niloticus*) exposed to concentrations of 1 and 10 mg/L small (10-30 nm) and large (100 nm) ZnO-NPs for 14 days, GSH in the gills increased statistically significantly in 1 and 10 mg/L small volume suspensions of ZnO-NPs, but in the intestine, it was found that GSH levels increased significantly on day 7 of exposure in 1 mg/L small suspension and 10 mg/L large suspension of ZnO- NPs, on day 14, levels increased GSH in 10 mg/L suspensions as well as small ZnO-NPs, but decreased in 1 mg/L large ZnO NPs suspensions.

Hamed and Osman (2017) found an increase in GSH in African catfish (*C. gariepinus*) after being exposed to carbofuran, the increase in GSH. Also, **Abd El-Atti *et al.*, (2020)** found that crayfish (*P. clarkii*) exposed to 125 and 250 ppm Ag-NPs levels after 28 days had, an increase in GSH but no change in 25 ppm Ag-NPs.

Superoxide dismutase (SOD):

In (2013), Zhao *et al.*, found that for 144 h after fertilization (hpf), zebrafish embryos were exposed to 1, 5, 10, 20, 50, and 100 mg/L ZnO-NP, SOD Significantly increase except for 100 mg/L ZnO-NP where there was a significant decrease. Other results in **Suman *et al.*, (2015)** recorded that when *Chlorella vulgaris* was exposed to 50 mg/L, 100 mg/L, 200 mg/L, and 300 mg/L ZnO-NPs for 24 h and 72 h, (SOD) activity was significantly increased in *C. vulgaris* exposed to 200 mg/L and 300 mg/L ZnO-NPs

Asghar *et al.*, (2018) noted that SOD levels increased significantly when freshwater *catla catla* fish were exposed to ZnO- NPs (40 mg/ L) for 28 days, with an average diameter of 36.77 nm. Moreover, **Dekani *et al.*, (2018)** studied that when common carp (*Cyprinus carpio*) was exposed to ZnO-NPs at nominal concentrations from a diet of 0, 30, 100, and 500 mg kg-1 to the base diet, it was observed that the SOD that received 500 mg kg of the diet increased.

Abdel-Daim *et al.*, (2018) showed that a decrease in SOD is observed when Nile tilapia fish were exposed to ZnO-NPs (50 mg/L) for 30 days. Recently **Aziz *et al.*, (2020)** showed that when exposed *Labeo rohita* to ZnO-NPs for 40, 60, and 80 days with a percentage of 31.15 and 57.84 mg/L, and ZnO-NP was of size was 52.22 nm. a decrease in SOD was observed

Total antioxidant capacity (TAC):

The study conducted by **Thwala *et al.*, (2013)** noted that for 14 days, duckweed *Spirodela punctuta* was exposed to ZnO-NPs concentrations of 0.01, 0.1, 1, and 1000 mg/ L, a significant increase in TAC was observed at the lowest exposure concentration but samples at other (higher) concentrations remained similar to the control. In contrast, **Abbasalipourkabir *et al.*, (2015)** observed that for ten days, adult male Wistar rats were exposed. 50, 100 and 150, and 200 mg/kg ZnO-NPs, (TAC) decreased insignificantly in all groups when the concentration of ZnO-NPs was more. However, the group (200 mg/kg) showed a significant decrease in TAC compared to control.

Similarly, **Abass *et al.*, (2017)** stated that it was observed that exposure of young male Wistar albino rats to ZnO-NPs (350 mg/kg B.W) caused a significant decrease in TAC. And, **Yousef *et al.*, (2018)** showed that When mice were orally exposed each day to ZnO-NPs was 100 mg/kg BW for 75 days, a decrease in TAC activity was observed.

Khayal *et al.*, (2021) recorded decreased total tissue antioxidant capacity (TAC) when male albino rats were exposed to ZnO-NPs (85 mg/kg b wt/day; 6 days per week). In the same year, **Ibrahim *et al.*, (2021)** showed that day by day for 30 days, mice were injected with ZnO-NPs 100 and 500 mg/kg body weight, which significantly reduced levels (TAC).

Glutathione peroxidase activity (GPx):

Ali ***et al.*****, (2012)** showed that for 24 and 86 h, *Lymnaea luteola* (*L. luteola*) was exposed to ZnO-NPs (32 μg/ml) which reduced GPx. Other results by **Saddick** ***et al.*****, (2015)** found that When *O. niloticus* and Tilapia zillii were exposed to 500 and 2000 μg/ L Zn-NPs, fish exposed to 500 μg/L Zn-NPs showed a significant increase in GPx, but fish exposed to 2000 μg/ L Zn-NPs showed a significant decrease in GPx activity.

In **(2016)**, **Ansar** ***et al.*****,** noted that Some mice were exposed to ZnO-NPs (600 mg/kg body weight) for 7 days, and their GPx was reduced. This agrees with **Abdelazim** ***et al.*****, (2018)** reported that muscles were collected on the 7th and 15th days of the experiment, and a decrease in GPx was observed when the Nile tilapia were exposed to 1 and 2 mg/L ZnO-NPs.

Alkaladi (2018) observed that GPx was inhibited in liver and gill tissues when Nile tilapia were exposed to 1 or 2 mg/L of ZnO-NPs for 7 and 15 days. In the same year, **Yousef** ***et al.*****, (2018)** found that when mice were exposed to ZnO-NPs nanoparticles (~100 nm particle size), the dose of ZnO-NPs was 100 mg/kg body weight, and decreased GPx levels were observed.

Nitric Oxide enzyme activity (NO) assay:

Fahmy ***et al.*****, (2014)** recorded that for 3 weeks, the freshwater snail *Biomphalaria alexandrina* was exposed to ZnO-NPs (7 and 35 μg/ml, which induced a significant increment in NO activity in hemolymph and tissues. **Fahmy and Sayed (2017)** showed that When freshwater bivalve mollusk *Coelatura aegyptiaca* was exposed to ZnO-NPs (2, 10, and 50 mg/L) for 6 consecutive days. NO increased at the concentrations of ZnO-NPs (10 and 50 mg/L) significantly.

Wang *et al.*, (2017) noted that when exposed to rats at different doses ZnO-NPs (200, 400, and 800 μg /kg), 8 hours a day at 40 hours per workweek, NO was increased over a dose (400 μg /kg). In the same year, **Rani *et al.*, (2017)** found that for 6 weeksrats were exposed to ZnO-NPs (50 mg/kg body weight/rat), A comparison of NO values obtained in the kidney of ZnO-NPs and control rats, showed higher values though insignificant in the kidney of ZnO-NPs treated rats.

In **(2018), Esmaeilnejad *et al.*,** found that when parasitic worms *Haemonchus contortus* are exposed to ZnO-NPs (1, 4, 8, 12, and 16 ppm) by diluting the stock solution, so NO was increased with increasing concentrations, for 24 hours. According to **Koner *et al.*, (2018)** showed that when the facultative catfish (*Clarias magur*) was exposed to ZnO-NPs (5 and 10 μg/ml) for 48 hours, it stimulated and increased NO.

Malondialdehyde (MDA):

Xiong *et al.*, (2011) noted that when zebrafish were exposed to 5 mg/L ZnO-NPs for 7 days, an increase in the percentage of MDA in liver tissue was observed. In the same line, **Ali *et al.*, (2012)** showed that when the freshwater snail *Lymnaea luteola* (*L. luteola*) was exposed to ZnO-NPs (32 μg/ml), after exposure for 24 and 86 hours, an increase of MDA was observed.

In (2017), Taheri *et al.*, observed that Upon exposure to common carp (*Cyprinus carpio*), 10 mg/kg and 15 mg/kg ZnO-NPs orally caused significant increases and MDA levels. Other results by **Abdelazim *et al.,* (2018)** found that muscles were collected on the 7th and 15th days of the experiment, and an increase in MDA was observed when the Nile tilapia were exposed to 1 and 2 mg/L ZnO-NPs.

Moreover, **Abdel-Daim *et al.,* (2018)** showed that an increase in MDA is observed when Nile tilapia fish were exposed to ZnO-NPs (50 mg/L) for 30 days. Recently, **Mahboub *et al.,* (2022)** found that when Common Carp (*Cyprinus Carpio*) was exposed to a 13 mg/kg or 26 mg/kg diet for 30 days, MDA increased significantly with increasing concentrations.

Evaluation of serum immunoglobulin:

A study conducted by **Jacobsen *et al.,* (2015)** found that mice were exposed to ZnO-NPs (15 μg/animal) bolus doses for 24 hours, and IgM was increased. But, **Beitsayah *et al.,* (2018)** showed that for 21 days, common carp (*Cyprinus carpio*) was fed with ZnO-NPs of 0, 5, 10, and 15 mg/kg of feed, there were no significant changes in total immunoglobulin activity.

According to **Deore *et al.,* (2021)** observed that for 28 days, IgM and IgG increased when Wistar rats were orally exposed to ZnO-NPs (100 mg/kg, body weight). But results by **Mahmoud *et al.,* (2021)** showed that when broiler chicks are exposed to ZnO-NPs (10, 20, or 40 mg/kg) for 42 days, not show any significant differences in IgM and IgG

Recently, **Kargin (2021)** recorded that When mice were exposed for 14 days. Three sub-lethal dose groups of ZnO-NPs (1, 5, and 25 mg/kg/day) were used. A decrease in IgM and IgG is observed. In the same year, **Rashidian *et al.,* (2021)** found that when common carp (*Cyprinus carpio*) was exposed to ZnO-NPs (58.85 mg/L), a decrease in total immunoglobulin (Total Ig) content was observed.

Blood parameters:

Crestani *et al.*, (2006) showed that when silver catfish, *R. quelen* was exposed to clomazone herbicide (0.5 and 1.0 mg/L) for 86 and 182h, Ht and Hb were decreased But, **Nieves-puigdoller *et al.*, (2007)** found that when Atlantic salmon, *S. salar* L. exposed to atrazine (100 μg/l) for 21 days, noted Ht was increasing.

In (2017), Chahardeh Baladehi *et al.*, found that When *Oncorhynchus mykiss* was exposed to ZnO-NPs (40 and 80 mg/ L), RBC and HB showed a significant decrease, while there was no significant difference in MCHC and showed a significant increase in types of WBCs after 5 days exposure to ZnO-NPs. **Suganthi *et al.*, (2018)** showed that When freshwater fish *O. mossambicus* was exposed to different behavior patterns with concentrations of 30, 50, and 70 ppm of ZnO-NPs, significantly reduced number of RBC, Hb and Ht was observed. Red blood indicators such as MCV, MCH, MCHC, and oxygen tolerance. A decrease in leukocytes was observed.

Yahya *et al.*, (2018) noted that mice were orally administered with ZnO-NPs (10 mg/kg/day) for 28 days, the concentration of erythrocytes, hemoglobin, and HCT was significantly lower, while a significant increase in MCV and a decrease in MCH values and no specific changes were observed in MCHC values. Recently, **Khan *et al.*, (2022)** showed that for 86 hours, A very high effect was recorded in the hematological parameters, which showed a significant ($p < 0.05$) increase in the number of white blood cells and platelets when *O. mossambicus* was exposed to ZnO-NPs (20 ppb) where the glass aquarium containing 40 L water.

ZnO-NPs and histopathological changes:

Hao *et al.*, (2013) noted that when juvenile carp (*Cyprinus carpio*) was exposed to ZnO-NPs (50 mg/L), some histological changes were observed such as atrophied and damaged cell membrane, swelling, and distorted organelles and the direction of the total disorder of the gill cells; In the liver were observed apoptosis, cell membrane rupture, hepatocyte desquamation, even severe cellular necrosis. Also, **Chupania *et al.*, (2018)** report that common carp (*Cyprinus carpio* L.) were fed with doses of 50 and 500 mg of ZnO-NPs per kg of feed for 6 weeks. Fish samples were taken every two weeks. Some histopathological changes appeared, such as some lamellar batons, a few aneurysms in the gills, and the appearance of some histological changes such as bleeding, and congestion. Mild tubular necrosis and tubular degeneration) in whole fish after two weeks of exposure to ZnO-NP.

According to **Shahzad *et al.*, (2018)** showed when tilapia was exposed to ZnO-NPs (0.5, 1.0, 1.5 mg/L) for 14 days, thickening of primary and secondary gill plates was observed. In the liver, necrosis, and apoptosis with condensed nuclei Clustered bodies and nuclei were observed. **Harsij *et al.*, (2021)** found that When goldfish (*Carassius auratus*) were exposed to zinc oxide, exposed to 20 and 50 mg/L of ZnO-NPs for 7 and 14 days, histological changes were observed in the gills such as destruction of the secondary lamina, hyperplasia of the epithelium, necrosis and filament epithelium proliferation, fusion of the secondary lamellae; While in the liver cell enlargement, oedema, Obstruction of blood flow and intercellular edema, degeneration and necrosis in the liver tissue.

Recently, **Khan *et al., (*2022)** observed that in the gills there are irregularities of the gill plate, rupture of the cartilaginous pulp, and hyperemia; and in the liver disintegration of the plasma membrane of the liver, ulceration, necrosis, enlargement, and formation of vacuoles; In the intestine, villi shortening, necrosis, separation and fusion of villi and severe goblet cell formation were observed when tilapia (*O. mossambicus*) was exposed to ZnO-NPs (20ppb) for 86 hours. In the same year, **Rajkumar *et al.*, (2022)** recorded those histological changes in gills, liver, and muscles such as platelet fusion, blood vessel dilation, cytoplasmic discharge, nuclear change, necrotic muscle fibers, and swollen nuclei were also revealed when freshwater fish (*Cyprinus carpio*) was exposed to various concentrations of ZnO-NPs (0.382, 0.573 and 1.146 mg/ L).

Parsley (*petroselinum*):

Farzaei *et al.*, (2013) recorded that parsley is one of the Umbelliferae family, originated in the Mediterranean and is currently cultivated in several places around the world, and is used in the treatment of digestive system, heart diseases, urinary tract, skin diseases, high blood pressure, hepatotoxicity, and diabetes. **Rahmat *et al.*, (2018)** Parsley is a leafy vegetable, its name is in Greek (*petroselinum*), and it is distinguished from other vegetables by its distinctive smell.

Önenç and Açıkgöz (2005); Altunbaş and Türel (2009) reported that parsley is used in traditional farming Medicines. **Zhang *et al.*, (2006); Altunbaş and Türel (2009)** studied that green and dried parsley leaves can be added to food as spices, spices, and food additives. **Charles (2012)** demonstrated that it is used in cosmetics.

Clark and Menary (1983) found that parsley leaves contain α-pinene (22.71%), beta-pinene (17.15%), myrcene (5.08%), beta-phyllandrin (11.57%), Terbinolin (3.03%), 1,3,8-P-Mentathrine (21.05%), P-Cymene (2.24%), Myristicin (4.86%) and apiole (7.48%). According that **MacLeod *et al.*, (1985)** showed that 4-isopropenyl-1-methylbenzene, apiole β-phellandrene, and 1,3,8-p-menthatriene are the main compound that gives parsley leaves a distinctive aroma. Moreover, **Pokorny and Pánek (2012)** noted that it contains antioxidants such as Among them are flavones (apigenin and luteolin), xanthophylls (Lutein and zeaxanthin), as well as some components of the essential oil, such as apiol and myristicin, can be distinguished (Bocornio).

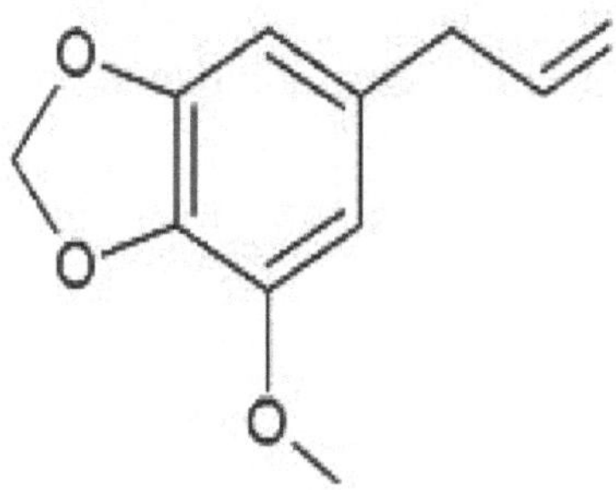

(1) Myristicin

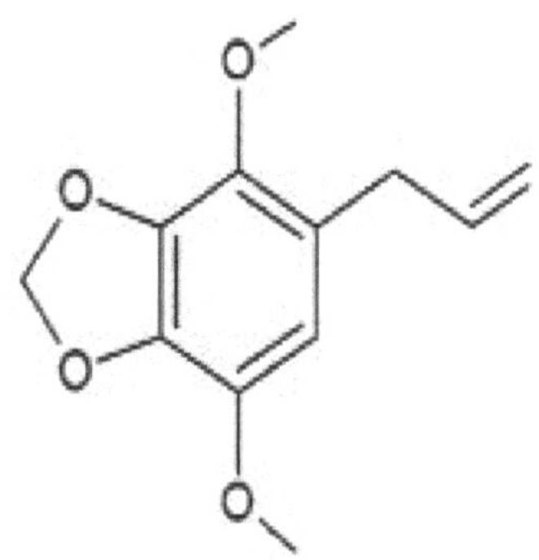

(2) Apiol

Fig (1): Chemical structures of the major constituents of parsley:

(1) myristicin (2) apiole

Other results showed that **Huda *et al.*, (2014)** found that parsley leaves are an important source of vitamins and minerals such as vitamins C, A, and E; Iron, calcium, phosphorous, and manganese are also used for medicinal purposes, so it is given during urinary tract infections because it is a diuretic. **Hassan *et al.*, (2004); Abdul Latif *et al.*, (2017)** noted that parsley also contains antioxidants, anti-inflammatory and immune stimulants. The results are suitable to **Lefsrud *et al.*, (2007); Fu *et al.* (2012)** who investigated that parsley is one of the sources of lutein, which has a scavenging activity against (ROS). Also, **Wong and Kitts (2006)** recorded that parsley contains polyphenols, which are an important source of antioxidants.

Material and
Methods

MATERIAL AND METHODS

Material:

Experimental fish:

In the present study, 400 Nile tilapia (*Oreochromis niloticus)* were obtained from alqanatir alkhayria, El qalyubia governorate, private fish farm, Egypt. The average body weight was 18 ± 2 g. The fish were healthy and free from any visible lesions. Fish were transported to the fish disease laboratory, at the Department of Aquatic Animals medicine Faculty of Veterinary Medicine, Benha University in large plastic containers supplemented with battery aerators as a source of oxygen. In the laboratory, fish were acclimatized in the glass aquaria for two weeks, before the experiments.

Management

Fish:

Glass aquaria measuring (60x47x32cm) (60)-liter capacity were used, and ten fish were kept in each glass aquarium containing dechlorinated tap water. The aquaria were cleaned and water was changed every 48 hours for the experimental and control group. The aquaria were supplied with electric air pumps where dissolved oxygen was maintained at 6±0.5 ppm. All aquaria pH ranged from 7.4 ±0.3 by regulator water change, water temperature was adjusted at 28±2°C

Chemicals:

1- Dried parsley leaves were obtained from Mashreq Company for Business Development - Smouha - Alexandria (Bab Al Sham)
2- Nano-dried parsley leaves were prepared at the Faculty of Nanotechnology, Cairo University. Egypt.

Other chemicals used:

Sodium thiosulphate: used for dechlorinization of tape water. Were purchased from Stanbio Company, Cairo, Egypt

Deionizing water to form a suspension of ZnO-NPs

Chemical kits: ALP, ALT, AST, urea, creatinine, uric acid, total protein, albumin, globulin, glucose, and cortisol, were purchased from Stanbio Company, Cairo, Egypt.

Chemicals for antioxidant determination (LPO, GSH, CAT, SOD, and TAC) were bought from a Bio-diagnostic Company, in Cairo, Egypt.

Serum IgM, and IgG levels were determined by the protocol of ELISA kits **(Cusabio Biotech Co. Ldt, USA)**

Methods:

Synthesis of ZnO nanoparticles (ZnO-NPs)

In a normal process, zinc acetate dihydrate (Zn $(CH_3COO)2$•2 H2O, Loba Chemie, Mumbai, India) as a precursor, and an ammonia solution of 30–33% (NH3) in an aqueous solution (NH_4OH, Advent chembio, Mumbai, India) as a reducing agent, had been used. The ZnO nanoparticles were produced using dissolving the right amount of zinc acetate in 100 mL of deionized water to supply 0.1 M of a zinc ions solution. Subsequently, the zinc ions solution turned into subjected to ultrasonic wave irradiation using a Hielscher UP400S (four hundred W, 24 kHz, Berlin, Germany) at an amplitude of seventy-nine% and a cycle of 0.76 for five min at a temperature of forty ◦C. Then, the ammonia solution became introduced dropwise to the zinc ions solution beneath the impact of the ultrasonic waves, until the precipitation of ZnO-NPs was completed according to **(Ismail *et al.*, 2021).**

Synthesis of *Petroselinum crispum* nanoparticles (*P. crispum*-NPs)

Petroselinum crispum nanoparticles (*P. crispum*-NPs) were synthesized using the bottom-up process for synthesis nanoparticles according to Karthik et al. (2016). Where, 5 g of *P. crispum* in micron size were put in stainless steel dram of ball mill instruments (photon ball mill model of 210 s, Egypt.) with ceramics balls of 0.1 cm diameter and 30 g weight, and milling with vacuum condition at 1200 rpm. for 3 h.

Characterization of ZnO-NPs and *P. crispum* – NPs:

Characterization is classified into three branches namely composition, morphology, and index branch. The morphology, composition, charge, and particle size distribution of synthetic nanoparticles were assessed by **Ismail *et al.*, (2020)**, (Fig. 2). The purpose of characterization was to determine synthetic nanoparticles' properties where any small change in size or shape of nanoparticles gives us dramatic effect in bioactivity and toxicity. The morphology branch is used to identify the shape and texture of synthetic nanoparticles. However, Transmission Electron Microscopy (TEM) is used to archive this branch. TEM instrument manufactures the JEOL company model of JEM-2100 high-resolution. Preparation of samples before study under TEM must be done. Firstly, adding synthetic nanoparticles to double deionized water and sonication for 10 min using an ultrasound prop with 60 kHz, at the amplitude of 60% and 0.58 of a cycle (Up 400 s, Hielscher, German). Finally, 5–10 μm of desperation was put onto a carbon-coated copper grid. The composition branch is used to illustrate formed of synthetic nanoparticles without any secondary chemicals from the synthesis method. However, x-ray diffraction is used to determine the formation of synthetic without any second products. XRD instrument manufactured by Thermo Scientific company model of EQUINOX 1000. The light source was Cu Kα radiation with a current of 30 mA and an applied voltage of 38 kV. The 2θ angles ranged from 25◦ to 80◦ for ZnO nanoparticles and 5–80◦ for *P. crispum* with a scan speed of 0.33◦/ min Index branch used to determine charge (zeta potential) and size (DLS). However, Dynamic light scattering

(DLS) and Zeta potential determine by zeta seizer instrument manufacturer by the Malvern Company model of Nano Sight NS500.

XRD pattern illustrates the synthesis of ZnO of hexagonal lattice according to the database of Burcker database (COD 2107058) (Fig. 3). XRD pattern of *P. crispum* illustrates the amorphous nature (Fig. 4).

TEM images of ZnO and *P.crispum* nanoparticles illustrate subspherical shapes with clear edges without any agglomeration (Fig. 5 & 6).

Dynamic light scattering (DLS). Results of dynamic light scattering for ZnO and *P. crispum* nanoparticles that illustrate the size distribution were 25–35 nm for ZnO-NPs and 20–25 nm for *Petroselinum crispum* nanoparticles. Zeta potential illustrates − 8.2 mV for ZnO-NPs and − 30.2 mV for *P. crispum.* The high zeta value of *P.crispum* indicated good stability in an aqueous solution

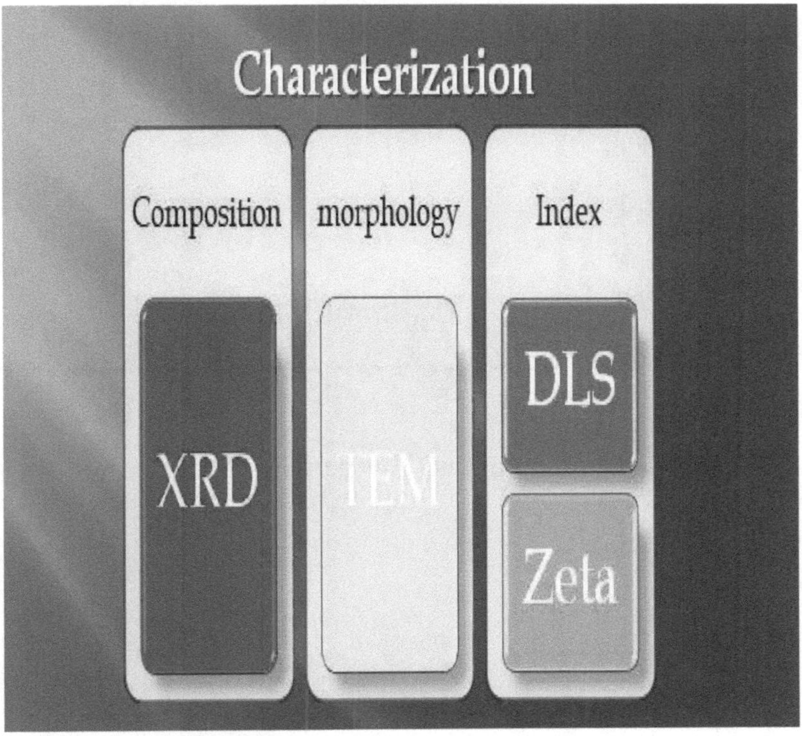

Fig. (2) Characterization of synthetic nanoparticles

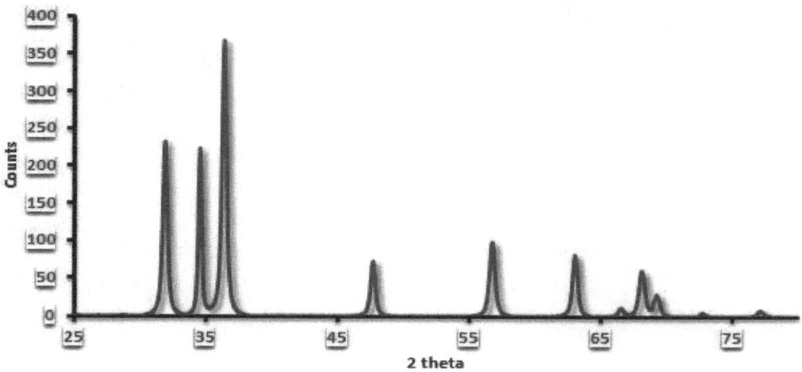

Fig. (3). XRD pattern of ZnO nanoparticles.

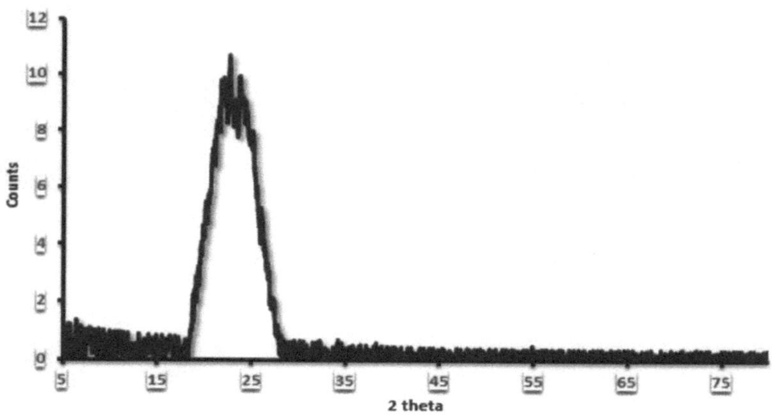

Fig. (4) XRD pattern of *Petroselinum crispum* nanoparticles.

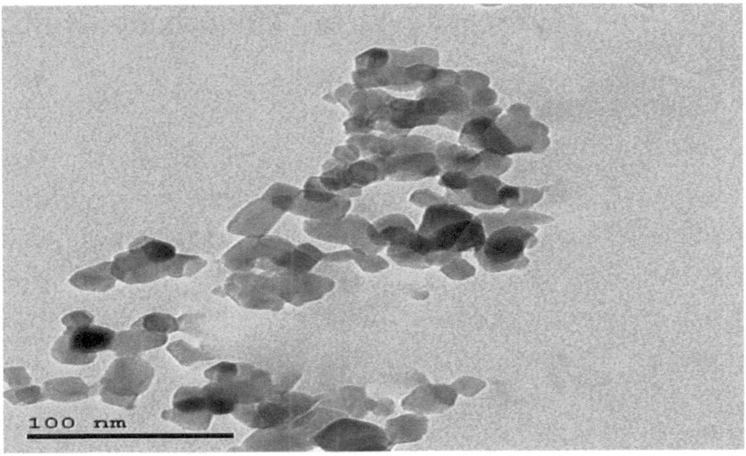

Fig. (5). TEM image of ZnO nanoparticles.

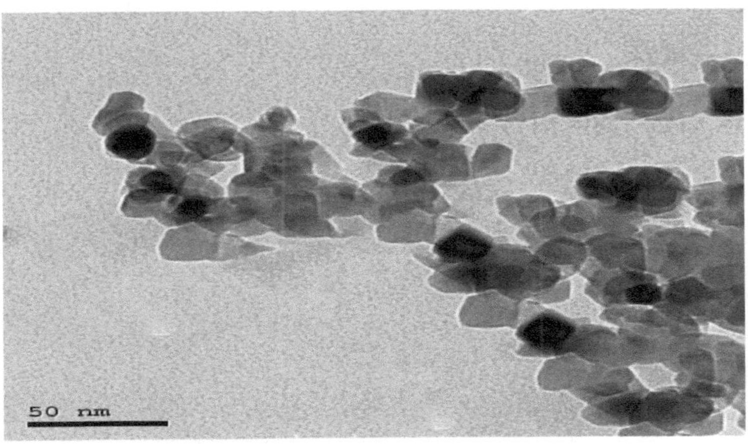

Fig. (6) TEM image of *Petroselinum crispum* (parsley) nanoparticles.

Determination of 96 hours half lethal concentration (LC_{50}) of zinc oxide nanoparticles:

The percentage of mortality was calculated and the values were transformed into a probit scale and analyzed as presented in Table 1. Regression lines of probit against the logarithmic transformation of concentrations were obtained. The slope function (S) and confidential limits (upper and lower) of the regression line with the Chi-square test were calculated. However, the LC50 value of ZnO-NPs was calculated by probit analysis (SPSS software), according to **Bhavya *et al.*, (2016)**

Table 1: Mortality of Nile tilapia at 96 h after treatment with different concentrations of ZnO-NPs.

Conc. ZnO-NPs (mg/l)	No. of Fishes	NO. of Mortality
0	10	0
10	10	1
20	10	2
40	10	2
80	10	4
160	10	8

Preparation diet:

Fish were fed twice daily on a balanced fish diet Prepared according to **Craig *et al.*, (2017).**

Dried parsley was purchased from the local market, turned into powder, and mixed with feed with a ratio of 4 g/kg diet. Nano parsley was prepared in the Faculty of Nanotechnology, Cairo University. Egypt. Batches of feed were left for 24 hours for air drying and stored at 4°c for day use **(AVMA, 2017)**

Experimental design:

One hundred eighty Nile tilapia were divided into 6 groups, and each group consisted of 3 replicates of 10 fish per aquarium and fed on a diet that was given to fish up to satiation daily at 8:00 and 14:00 for 6 weeks. Group I: was served as the control group without any treatment. Group II: served as the control group and co-administrated with parsley at the rate of (4mg/kg diet) according to **El-Barbary and Mehrim (2008)**. Group III: served as the control group and co-administrated with nano parsley at the rate of (4mg/kg diet) according to **El-Barbary and Mehrim (2008)** Group IV: was exposed to (8 mg/L) ZnO-NPs in deionized water was prepared according to **Amin *et al.*, (2021)** Group V: was exposed to (8 mg/L) ZnO-NPs and fed on (4g/kg diet) from parsley according to **El-Barbary and Mehrim (2008)**. Group VI: was exposed to (8 mg/L) ZnO-NPs and fed on (4g/kg diet) from nano parsley. During the experimental period, any fish mortalities or abnormalities were recorded.

Clinical investigation and post-mortem examination:

The experimentally exposed fish were kept under proper observation during the period of the experiment for any clinical abnormalities, postmortem (PM) lesions, or deaths. according to the method described by **(Amlacher, 1970)**

Sampling of blood and serum:

After 6 weeks of exposure to ZnO-NPs :

1-Blood samples: were collected from the heart vessels using plastic syringes in dry sterilized vials. Allowed the blood to clot then samples were centrifuged, at 3000×g at 4 °C for 15 min. Serum collected and stored at -20 °C until used, also using a plastic syringe moistened with EDTA for measuring Red blood cells (RBCs), White blood cells (WBCs), differential leukocytic count, and hemoglobin (Hb) concentration.

2-Tissues samples: Liver samples were collected and homogenized in phosphate-buffered saline, then centrifuged at 3000 rpm for 15 minutes and maintain to homogeneity -20°C until used in antioxidant analysis, Somethere Liver samples were collected and fixed in neutral buffered formalin for histopathological examination.

Neutral buffered formalin was prepared according to **Carleton *et al.*, (1967)** for the fixation of histopathological specimens. It was prepared as follows:

Formaline	100 ml
Sodium phosphate dibasic	6.5ml
Sodium phosphate	4 ml
D.W	800ml

Determination of total erythrocytes and leukocytic count:

Blood elements (RBCs and WBCs) were counted according to **Kanaev (1985)** using Haemocytometer and (Schwa's) solution, which formed from a mixture of Solution A and Solution B, the Solution was prepared by mixing an equal volume of Solution A and Solution B.

Procedure:

Blood samples (0.5 ml) were taken and completed to mark 101 with the dilution fluid. Shaken gently and allowed to settle for 2-3 minutes, a drop of mixed fluid after expelling the first drop was put on Haemocytometer and allowed to settle under the cover slide on the counting chamber for 2-3 minutes. Erythrocytes were counted in five large squares in the central square, while Leukocytes were counted in four corner squares of the hemocytometer-counting chamber and the average numbers were taken.

Calculation:

RBCs= (50) (V) (X1) = /mm^3
WBCs= (10) (V) (X2)/4=/mm^3
Where: V=dilution factor.
X1=number of cells calculated in 5secondry squares.
X2=number of cells calculated in 4 corner squares.

Determination of haemoglobin concentration (g/dl):

Blood haemoglobin concentration (g/dl) was determined quantitatively, by colorimetric methods, using commercial colorimetric Kits (Diamond Diagnostic, Egypt), according to the method described by **Campbell (2004)**. The blood was treated with Drabkin's reagent, which contained potassium ferricyanide, potassium cyanide, and potassium dihydrogen phosphate. The ferricyanide reacts with blood-forming methemoglobin, which is converted to cyanmethemoglobin by potassium cyanide. The color obtained was measured spectrophotometrically at 546 nm wavelength. The concentration of blood haemoglobin was calculated from the following Equation:

Blood haemoglobin concentration= A×36.8=g/dl,

Where: A= Absorbance of sample

Determination of Differential leukocytic count (DLC):

DLC was applied according to **Stoskopf, (1983).** The fixed blood film was dipped in diluted Giemsa stain for 15-20 minutes then rinsed with distilled water and left to dry and examined by an oil immersion lens. One hundred cells of the blood smear were differentiating; count and the percentages of different leukocytes were calculated.

Blood Indices:

Were calculated according to **Gupta, (1977)**

1- Mean corpuscular volume (fentoliter):

$$MCV= \frac{P.C.V\%}{RBCs\ (10^6 / mm^3)} \times 10$$

2- Mean corpuscular Haemoglobin (Pg / Cell):

$$MCH= \frac{Hb\ (g / dl)}{RBCs\ (10^6 / mm^3)} \times 10$$

3- Mean corpuscular Haemoglobin concentration:

(g/ 100ml):

$$MCHC = \frac{Hb\ (g / dl)}{P.C.V\%} \times 10$$

Biochemical analysis:

Determination of Alkaline phosphatase (ALP):

Serum ALP level was estimated using a stanbio kit as described by **Tietz, (1984).**

Determination of Serum amino transferases:

Serum ALT and AST levels were estimated using a stanbio kit as described by **Young, (1990).**

Determination of Serum urea:

Serum urea level was estimated using a stanbio kit as described by **Tietz, (1986).**

Determination of Serum creatinine:

Serum creatinine level was estimated using a stanbio kit as described by **Tietz and Norbert (1985).**

Determination of Serum total protein:

Serum total protein level was estimated using a stanbio kit as described by **Orsonneau *et al.*, (1989).**

Determination of Serum albumin:

Serum albumin level was estimated using a stanbio kit as described by **Dumas and Biggs (1972).**

Determination of Serum globulin:

It was calculated as the difference between the total protein and albumin.

Determination of Albumin /Globulin ratio:

It was calculated from the albumin present in the serum with the amount of globulin.

Determination of Serum glucose:

Glucose level was estimated using a stanbio kit as described by **Young, (1990).**

Determination of Serum cortisol:

Cortisol determined using RIA kit according to the method described by **Foster and Dunn (1974).** It was measured by using DSL-2100 ACTIVE®.

Serum lipogram analysis:

Determination of Cholesterol:

It was estimated as described by **Stein, (1986).**

Determination of Triglycerides:

It was determined according to **Scheletter and Nussel (1975)**

Oxidative stress and antioxidant enzymes:

The tissue of the liver were homogenized in a potter-Elvejhem glass/Teflon homogenizer and centrifuged then the supernatant was collected for determined the following parameters:

Determination of Reduced glutathione (GSH):

It was estimated according to the method described by **Beutler *et al.*, (1963).** The method was based on the reduction of 5.5 -dithiobis (2-nirtobenzoic acid) (DTNB) with glutathione (GSH) to produce a yellow compound. The reduced chromogen is directly proportional to (GSH) concentration and its absorbance could be measured at 405 nm.

Determination of Catalase enzyme (CAT):

According to the method described by **Aebi, (1984)** the concentration of catalase enzyme was estimated by using a Bio-Diagnostic kit. In this test, catalase reaction with a known quantity of H2O2. The reaction is stopped after exactly one minute with a catalase inhibitor.

$$2H_2O_2 \xrightarrow{\text{Catalase}} 2H_2O + O_2$$

Determination of Superoxide dismutase (SOD):

The concentration of superoxide dismutase enzyme in liver tissues was measured calorimetrically according to the method described by **Nishikimi *et al.*, (1972)**. This assay relied on the ability of the enzyme to inhibit the phenazine methosulphate-mediated reduction of nirtoblue tetrazolium dye.

Determination of Total antioxidant capacity (TAC):

TAC was determined using a commercial kit (Antioxidant assay kit, cat. No. CS0780; Sigma-Aldrich, USA). Absorbance was read at 405 nm using a Multiscan EX microplate reader (lab systems, Helsinki, Finland). Under conditions of this assay, the volume of plasma or diluted plasma was 10 µl and TAC values of samples were expressed as an equivalent of the mmol concentration of a (±)-6-hydroxy-2,5,7,8-tetramethylchromane-2- carboxylic acid (Trolox) solution. Trolox standard curve at the range of 0.00-0.42 mM was prepared for the assay according to the method described by **Koracevic *et al.*, (2001)**

Nitric Oxide enzyme activity (NO) assay:

Nitric oxide was estimated according to **Rajaraman *et al.*, (1998)**, NO is synthesized in biological systems by the enzyme Nitric Oxide synthase (NOS). NOS is a remarkably complex enzyme that acts on molecular oxygen, arginine, and NADPH to produce NO, citrulline, and NADH+.

The Biodiagnostic Nitrate Assay Kit provides an accurate and convenient method for the measurement of endogenous nitrite concentration as an indicator of nitric oxide production in biological fluids. It depends on the addition of Griess Reagent which converts nitrite into a deep purple azo compound, photometric measurement of the absorbance due to this azo chromophore accurately determines NO_2 concentration

Glutathione peroxidase activity (GPx):

Glutathione peroxidase (GPx, EC 1.11.1.8) activities were determined by detecting the non-enzymatic utilization of GSH as the reacting substrate at an absorbance of 412 nm after incubation with DTNB according to the method of **Moin, (1986)**. Briefly, the reaction mixture contained 0.8 ml of 0.1M Tris-HCl buffer (pH 8.8) with 12 mM sodium azide, 6 mM EDTA, 0.2 ml of homogenate, 0.1 ml of 4.8 mM glutathione, and 0.1 ml of 20 mM t-butyl hydroperoxide. The contents were incubated at 37°C for 10 min. The reaction was stopped by 0.2 ml of 20% TCA, and centrifuged. The supernatant was assayed for glutathione content with Ellman's reagent (38.6 mg of DTNB in 10 ml of 1% sodium citrate). A unit of enzyme activity is defined as the amount of enzyme catalyzing the formation of

1 μmol of GSH per min, and the activity of GPx was calculated based on tissue protein concentration.

Malondialdehyde (MDA):

MDA level was measured after incubation at 85°C with thiobarbituric acid (TBA) in aerobic conditions (pH 3.4). The pink color produced by these reactions was measured spectrophotometrically at 532 nm according to **Ohkawa *et al.*, (1978).**

Evaluation of serum immunoglobulins (IgM+ IgG):

Serum IgM, and IgG levels in treated and control fish were measured spectrophotometrically following the protocol of ELISA kits **(Cusabio Biotech Co. Ldt, USA)** as the following:

50μl of Standard or Samples were added per well then 50μl of HRP-conjugate was added to each well then 50μl Antibody was added to each well. Mix well and then were incubated for one hour at 37°C. Each well was aspirated and washed with washing buffer (200μl); the process was repeated two times for a total of three washes. After the last wash, any remaining Wash Buffer was removed by aspirating or decanting. The plate was inverted and blotted against clean paper towels. 50μl of **Substrate A** and 50μl of **Substrate B** were added to each well then mixed well and incubated for 15 minutes at 37°C in a dark place. 50μl of Stop Solution was added to each well and mixed well. The optical density of each well was determined within 10 minutes by using a microplate reader set to 6 weeks0 nm. All samples and standards were assayed in duplicate.

Histopathological examination

At the end of the experimental period, tilapia was dissected and internal organs were carefully examined for any abnormal macroscopic changes. The small tissue specimens were collected from the liver of fish in all groups and immediately fixed in 10% neutral buffered formalin. After proper fixation, tissue specimens were washed, dehydrated in a gradually ascending series of ethyl alcohol, cleared in xylene, and embedded in paraffin wax. Tissue paraffin sections were sectioned with a rotary microtome and 5 µm sections were stained with H&E stain according to the methods described by **Suvarna *et al.*, (2018)** for microscopic examination of the hepatic parenchyma. The histopathological changes were recorded using Nikon's Eclipse E800 microscope and photomicrographs were taken using a digital camera.

Statistical Analysis:

The data were presented as means ± SE. Data were subjected to one-way ANOVA to evaluate the effects of ZnO-NPs toxicity, parsley, and nano parsley administration. followed by Duncan's post hoc test to compare between groups, $P<0.05$ was considered statistically significant. All the data analyses were done using **SPSS (2018)**.

Results

RESULTS

96 hours half lethal concentration (LC_{50}) of zinc oxide nanoparticles in *Oreochromis niloticus:*

In this study, the half-lethal concentration (LC_{50}) of ZnO-NPs is 80.9 mg/L. (Table 2). A sub-lethal ZnO-NPs concentration (1/10 LC_{50}) of 8 mg/L was used in this investigation.

Table (2): Estimated ZnO-NPs concentration values and confidence limits Estimated ZnO-NPs concentration values and confidence limits:

Probability	85% Confidence Limits for conc.		
	Estimate	Lower Bound	Upper Bound
0.01 (LC1)	3.168	0.046	8.682
0.05	8.186	0.481	18.303
0.1 (LC10)	13.576	1.635	26.3
0.2	25.052	6.761	43.454
0.3	38.867	16.738	70.164
0.4	56.838	30.788	124.588
0.5 (LC50)	80.884	46.814	247.888
0.6	115.103	64.833	540.471
0.7	167.888	87.661	1308.056
0.8	261.141	120.741	3786.26
0.8 (LC80)	481.887	183.443	17071.23
0.85	788.228	256.258	58742.32
0.88	1412.447	370.841	246228.8
0.88 (LC88)	2064.623	473.551	634459.6

Clinical signs and postmortem findings:

In this study, *O. niloticus* exposes to (8 mg/L) of ZnO-NPs for 6 weeks, showed abnormal skin pigmentation, moving towards the air pump, staying for a long time, a certain location, also irregular movement, loss of equilibrium, anorexia, detached scales, frayed fins, and Postmortem lesions showed lipid degeneration in the liver with paleness in other internal organs whereas observed. Whereas, co-administration with parsley leaves and parsley nanoparticles (4g/kg diet) in the groups (V and VI) reduced the abnormal behavior of *O.niloticus* exposed to ZnO-NPs.

Fig (7):

Fig. 7A: Showing detached scales, and frayed fins in *O.niloticus* exposed to ZnO-NPs (8 mg/ L.) **Fig 7B**: *O. niloticus* from group (I) (control) **Fig. 7C:** Showing lipid degeneration in the liver with paleness in other internal organs of *O.niloticus* exposed to ZnO-NPs (8 mg/ L.). **Fig 7D**:Anatomy of *O. niloticus* from group (I) (control)

Serum biochemical analysis:

<u>Serum liver enzymes:</u>

Serum Aspartate amino transferase (AST):

AST levels in serum samples of *O. niloticus* exposed to ZnO-NPs (8 mg/L) alone or co-treated with parsley leaves and parsley nanoparticles (4g/kg diet) for 6 weeks are shown in Table (3) and Fig. (8). The data recorded that the level of AST was significantly ($P<0.05$) increased at the end of the experiment in the group (IV). Meanwhile, exposure *O. niloticus* with ZnO-NPs (8 mg/L) and supplemented with parsley nanoparticles (4g/kg diet) reduced the level of AST to the normal level of the control group. The enhancement in AST levels was more evidence in group (VI) which supplemented with parsley nanoparticles.

Serum alanine amino tranferase (ALT):

ALT levels in serum samples of *O. niloticus* exposed to ZnO-NPs (8 mg/L) alone or co-treated with parsley leaves and parsley nanoparticles (4g/kg diet) for 6 weeks are shown in Table (3) and Fig. (8). The data recorded that the level of ALT was significantly ($P<0.05$) increased at the end of the experiment in the group (IV). Meanwhile, exposure *O. niloticus* with ZnO-NPs (8 mg/L) and supplemented with parsley nanoparticles (4g/kg diet) reduced the level of ALT to the normal level of the control group. The enhancement in ALT levels was more evidence in group (VI) which supplemented with parsley nanoparticles.

Serum Alkaline phosphatase (ALP):

The mean values of ALP in serum samples of *O. niloticus* exposed to ZnO-NPs (8 mg/L) alone or co-administrated with parsley leaves and parsley nanoparticles (4g/kg diet) for 6 weeks are demonstrated in Table (3) and Fig. (8). The present results revealed that the level of ALP was significantly ($P< 0.05$) increased at the end of the experiment in group (IV). However, exposured *O. niloticus* to ZnO-NPs (8 mg/L) and supplemented with parsley leaves and parsley nanoparticles (4g/kg diet), reduced the level of ALP to the normal level of the control group.

Table (3): Effect of ZnO-NPs exposure (8 mg/L) on serum liver function in Nile tilapia, *O. niloticus*, and treatment with parsley leaves and parsley nanoparticles (4g/kg diet) for 6 weeks.

Parameters / Groups	AST (μ/l)	ALT (μ/l)	ALP (μ/l)
Group I (Control)	20.17 ± 0.88^{d}	65.38 ± 8.20^{d}	21.31 ± 2.37^{b}
Group II (Parsley)	38.63 ± 4.78^{b}	82.76 ± 3.67^{b}	23.48 ± 2.67^{b}
Group III (Nano parsley)	21.53 ± 1.31^{d}	58.86 ± 3.40^{d}	21 ± 1.43^{b}
Group IV (ZnO-NPs)	83.70 ± 2.65^{a}	126.23 ± 5.55^{a}	48.86 ± 4.42^{a}
Group V (ZnO-NPs +Parsley)	32.31 ± 4.70^{bc}	81.52 ± 4.18^{bc}	25.18 ± 2.11^{b}
Group VI (ZnO-NPs +Nano parsley)	25.25 ± 1.78^{cd}	71.08 ± 3.38^{cd}	25.62 ± 2.54^{b}

Data were founded as mean ± SE values in the same column for each group were (P<0.05).

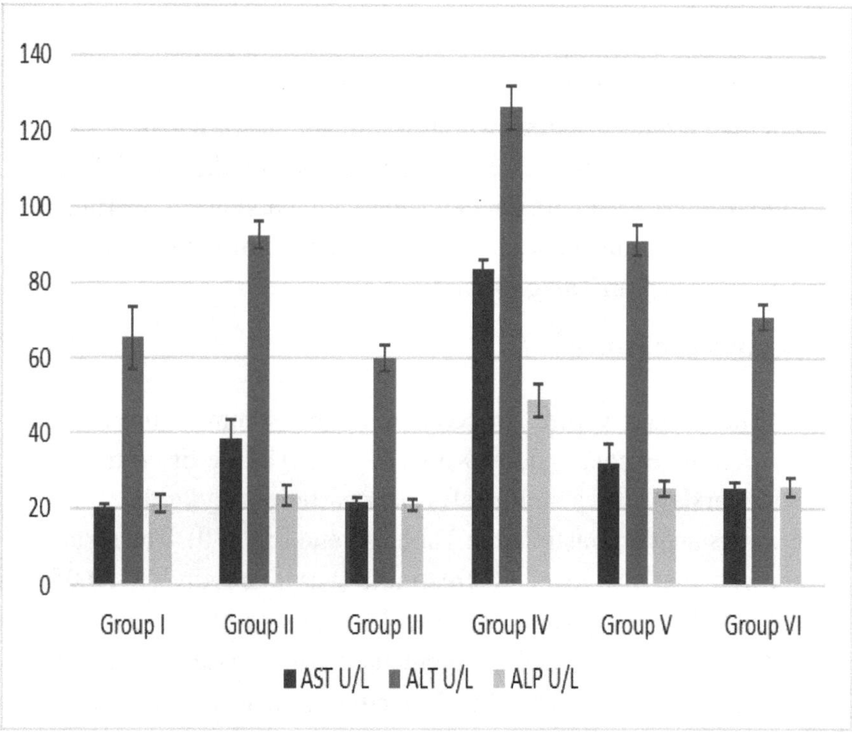

Fig. (8): Histogram showing the variations in AST (U/L), ALT (U/L), and ALP (U/L) levels of *O. niloticus* exposed to ZnO-NPs (8 mg/L) and treated with parsley leaves and parsley nanoparticles (4g/kg diet) 6 weeks.

Serum kidney function:

Serum urea:

The present results show that after 6 weeks of exposure to ZnO-NPs (8 mg/L), there was a significant increase in urea level in group (IV). However, exposed Nile tilapia with ZnO-NPs (8 mg/L) and treated with parsley leaves and parsley nanoparticles (4g/kg diet) reduced the level of urea to the normal level of the control group, in Table (4) and Fig. (9).

Serum creatinine:

The mean values of creatinine in serum samples of *O.niloticus* exposed to ZnO-NPs (8 mg/L) alone or co-treated with parsley leaves and parsley nanoparticles (4g/kg diet) after 6 weeks are demonstrated in Table (4) and Fig. (10). The results showed that the level of creatinine was markedly ($P < 0.05$) increased in group (IV). Meanwhile, exposing *O.niloticus* with ZnO-NPs (8 mg/L) and supplemented with parsley leaves and parsley nanoparticles (4g/kg diet) reduced the level of creatinine to the normal level of the control group.

Table (4): Effect of ZnO-NPs (8 mg/L) on serum kidney function in Nile tilapia, *O. niloticus* and treated with parsley leaves and parsley nanoparticles (4g/kg diet) for 6 weeks.

Parameters / Groups	Urea (mg/dl)	Creatinine(mg/dl)
Group I (Control)	18.16 ± 1.83^{c}	0.21 ± 0.02^{b}
Group II (Parsley)	24.08 ± 1.41^{bc}	0.25 ± 0.06^{b}
Group III (Nano parsley)	20.86 ± 1.88^{bc}	0.25 ± 0.06^{b}
Group IV (ZnO-NPs)	41.46 ± 2.15^{a}	0.7 ± 0.07^{a}
Group V (ZnO-NPs +Parsley)	25.22 ± 0.55^{b}	0.23 ± 0.02^{b}
Group VI (ZnO-NPs +Nano parsley)	20.02 ± 1.07^{c}	0.17 ± 0.02^{b}

Data were founded as mean ± SE values in the same column for each group were (P<0.05).

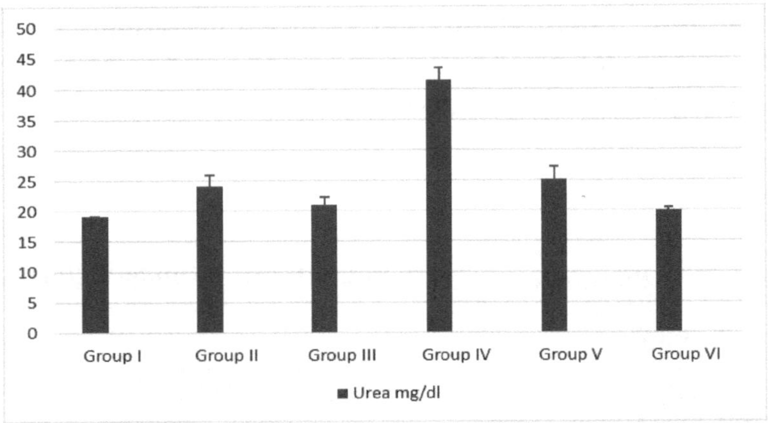

Fig. (9): Histogram showing the variations in urea (mg/dl) levels of *O. niloticus* exposed to ZnO-NPs (8 mg/L) and treated with parsley leaves and parsley nanoparticles (4g/kg diet) 6 weeks.

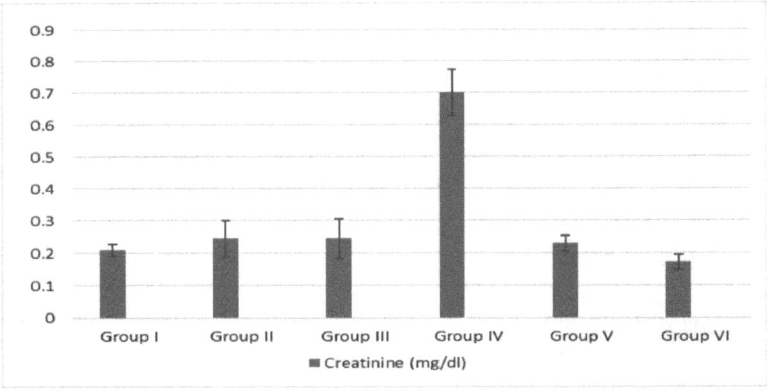

Fig. (10): Histogram showing the variations in creatinine (mg/dl), levels of *O.niloticus* exposed to ZnO-NPs (8 mg/L) and treated with parsley leaves and parsley nanoparticles (4g/kg diet) 6 weeks.

Serum Immune function:

Serum total protein

After 6 weeks of exposure values of total proteins in serum samples of *O. niloticus* exposed to ZnO-NPs (8 mg/L) alone or treated with parsley leaves and parsley nanoparticles (4g/kg diet) are shown in Table (5) and Fig. (11). The statistical analysis recorded that the level of total proteins was significantly ($P<0.05$) increased in group (IV). Meanwhile, exposed *O. niloticus* with ZnO-NPs (8mg/L) and treated with parsley leaves and parsley nanoparticles (4g/kg diet) reduced the level of total proteins in the control group.

Serum albumin:

From Table (5) and Fig. (11), serum albumin showed markedly ($P< 0.05$) increased after 6 weeks in the group (IV). However, exposed Nile tilapia with ZnO-NPs (8 mg/L) and supplemented with parsley leaves and parsley nanoparticles (4g/kg diet) reduced the level of albumin to the normal level of the control group.

Serum globulin:

Globulin in serum samples of Nile tilapia exposed to (8 mg/L) ZnO-NPs alone or co-treated with parsley leaves and parsley nanoparticles (4g/kg diet) for 6 weeks is demonstrated in Table (5) and Fig. (11). The statistical analysis recorded that the level of globulin was markedly ($P<0.05$) increased in group (IV). However, exposure of Nile tilapia with ZnO-NPs (8 mg/L) and treatment with parsley leaves and parsley nanoparticles (4g/kg diet) reduced the level of globulin to the normal level of the control group.

Serum A/G ratio:

The results presented in Table (5) and Fig. (11) showed a significant ($P< 0.05$) decrease in the level of A/G ratio after 6 weeks in the group (IV). Meanwhile, exposure of *O. niloticus* with ZnO-NPs (8 mg/L) and treatment with parsley leaves (4g/kg diet) increase the level of the A/G ratio to the control group. The results were more evident in group (V) treated with parsley than in group (VI) treated with parsley nanoparticles.

Table (5): Effect of ZnO-NPs (8 mg/L) on serum Total protein, albumin, globulin, and A/G ratio function in Nile tilapia, *O. niloticus* and treated with parsley leaves and parsley nanoparticles (4g/kg diet) for 6 weeks.

Parameters / Groups	Total protein (g/dl)	Albumin (g/dl)	Globulin (g/l)	A/G ratio
Group I (Control)	4.33 ± 0.24^{bc}	2.3 ± 0.08^{ab}	2.03 ± 0.24^{bc}	1.18 ± 0.18^{ab}
Group II (Parsley)	3.35 ± 0.25^{d}	1.88 ± 0.14^{c}	1.46 ± 0.14^{c}	1.31 ± 0.12^{ab}
Group III (Nano parsley)	3.56 ± 0.21^{d}	2.16 ± 0.05^{bc}	1.4 ± 0.26^{c}	1.72 ± 0.34^{a}
Group IV (ZnO-NPs)	5.30 ± 0.3^{a}	2.51 ± 0.08^{a}	2.85 ± 0.28^{a}	0.88 ± 0.05^{b}
Group V (ZnO-NPs +Parsley)	3.68 ± 0.17^{cd}	2.33 ± 0.10^{ab}	1.38 ± 0.2^{c}	1.87 ± 0.35^{a}
Group VI (ZnO-NPs +Nano parsley)	4.8 ± 0.15^{ab}	2.25 ± 0.04^{ab}	2.56 ± 0.14^{ab}	0.83 ± 0.08^{b}

Data were founded as mean ± SE values in the same column for each group were ($P<0.05$).

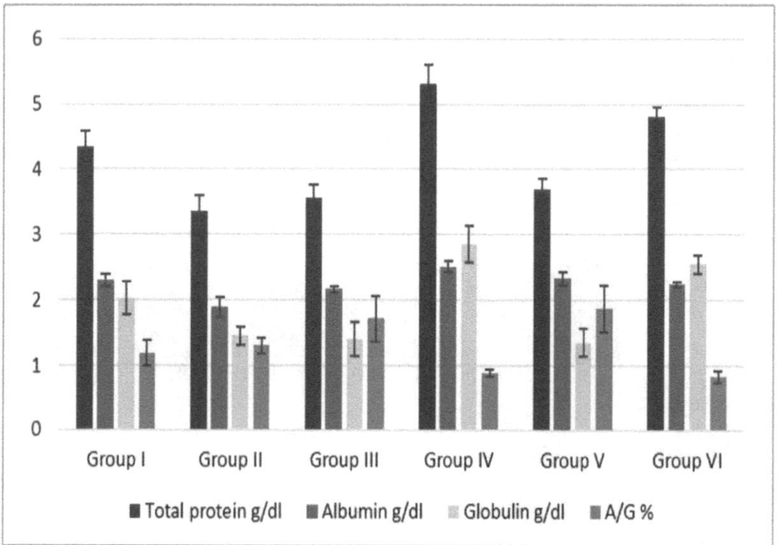

Fig. (11): Histogram showing the variations in Total protein (g/dl), Albumin (g/dl), Globulin (g/dl), and A/G ratio levels of *O. niloticus* exposed to ZnO-NPs (8 mg/L) and treated with parsley leaves and parsley nanoparticles (4g/kg diet) 6 weeks.

Serum stress indicators:

Serum glucose:

The mean values of glucose in serum samples of *O. niloticus* exposed to ZnO-NPs (8 mg/L) alone or co-administrated with parsley leaves and parsley nanoparticles (4g/kg diet) for 6 weeks were demonstrated in Table (6) and Fig. (12). Level of glucose in serum samples was increased significantly (P<0.05) after 6 weeks of ZnO-NPs exposure in comparison with the values of the control group. However, parsley leaves and parsley nanoparticles (4g/kg diet) reduced glucose levels to nearly the normal level, mainly in fish that received parsley nanoparticles (4g/kg diet) in group (VI).

Serum cortisol:

The mean values of cortisol in serum samples of *O. niloticus* exposed to ZnO-NPs (8 mg/L) alone or co-administrated with parsley leaves and parsley nanoparticles (4g/kg diet) for 6 weeks were demonstrated in Table (6) and Fig. (12). The statistical analysis showed that the cortisol level was significantly (P<0.05) increased after 6 weeks of ZnO-NPs exposure as compared to the control group. Alternatively, treatment of *O. niloticus* with parsley leaves and parsley nanoparticles resuted in a significant reduction in cortisol concentrations. Especially in group (VI) which received parsley nanoparticles (4g/kg diet) in comparison with Nile tilapia exposed to ZnO-NPs alone group (IV)

Table (6): Effect of ZnO-NPs (8 mg/L) on serum Glucose and cortisol of Nile tilapia, *O. niloticus* and treated with parsley leaves and parsley nanoparticles (4g/kg diet) for 6 weeks.

Parameters / Groups	Glucose (mg/dl)	Cortisol (µg/dl)
Group I (Control)	88.38 ± 5.54^{c}	18.51 ± 0.56^{c}
Group II (Parsley)	147.21 ± 3.85^{a}	41.2 ± 3.67^{b}
Group III (Nano parsley)	100.16 ± 5.4^{c}	22.47 ± 1.35^{c}
Group IV (ZnO-NPs)	141.86 ± 8.58^{a}	58.34 ± 7.76^{a}
Group V (ZnO-NPs +Parsley)	124.33 ± 3.88^{b}	44.73 ± 2.06^{b}
Group VI (ZnO-NPs +Nano parsley)	71.73 ± 2.85^{d}	25.74 ± 2.0^{c}

Data were founded as mean ± SE values in the same column for each group were ($P<0.05$).

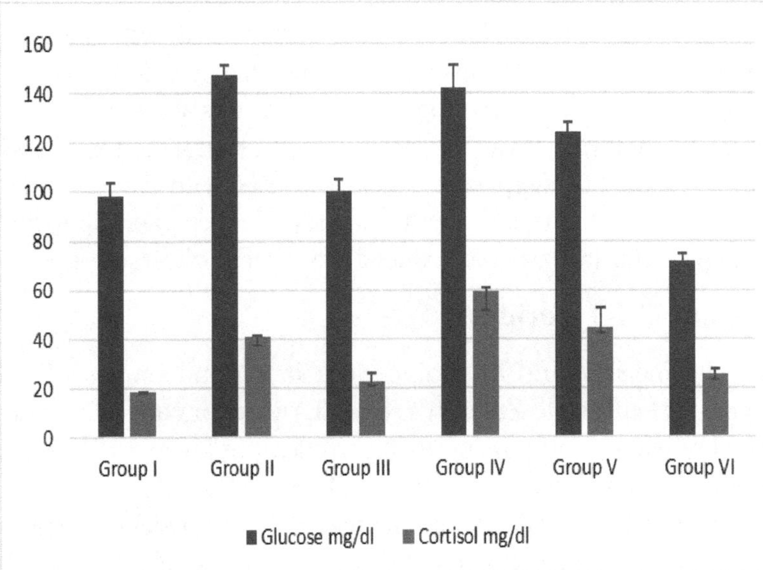

Fig. (12) Histogram showing variations in glucose (mg/dl), and cortisol hormone (mg/dl) levels of *O. niloticus* exposed to ZnO-NPs (8 mg/L) and treated parsley leaves and parsley nanoparticles (4g/kg diet) 6 weeks.

Serum Total lipids:

Serum Cholesterol:

Cholesterol in serum samples of *O. niloticus* exposed to ZnO-NPs (8 mg/L) alone or co-administrated with parsley leaves and parsley nanoparticles (4g/kg diet) for 6 weeks are recorded in Table (7) and Fig. (13). The results showed that the level of cholesterol was markedly ($P<0.05$) increased in group (IV). Meanwhile, exposure of *O. niloticus* with ZnO-NPs (8 mg/L) and treatment with parsley leaves and parsley nanoparticles (4g/kg diet) reduced the level of cholesterol.

Serum Triglycerides:

The mean values of triglycerides in serum samples of *O. niloticus* exposed to ZnO-NPs (8 mg/L) alone or co-treated with parsley leaves and parsley nanoparticles (4g/kg diet) 6 weeks were demonstrated in Table (7) and Fig. (13). Serum triglycerides showed marked decreased after 6 weeks of ZnO-NPs exposure in comparison with the values of the control group. Meanwhile, supplementation with parsley leaves and parsley nanoparticles (4g/kg diet) increased the level of triglycerides near the normal levels and this enhancement was more evident in fish that received parsley (4g/kg diet) for 6 weeks of exposure.

Table (7): Effect of ZnO-NPs (8 mg/L) on cholesterol and triglycerides of Nile tilapia, *O. niloticus* and treated with parsley leaves and parsley nanoparticles (4g/kg diet) for 6 weeks.

Parameters / Groups	Cholesterol (mg/dl)	Triglycerides (mg/l)
Group I (Control)	48.08 ± 3.65^{d}	72.84 ± 7.42^{ab}
Group II (Parsley)	50.86 ± 3.74^{d}	70.85 ± 8.31^{ab}
Group III (Nano parsley)	55.26 ± 4.03^{cd}	62.78 ± 2.71^{ab}
Group IV (ZnO-NPs)	84.8 ± 3.5^{a}	58.26 ± 3.78^{b}
Group V (ZnO-NPs +Parsley)	68.67 ± 3.84^{b}	78.75 ± 2.46^{a}
Group VI (ZnO-NPs +Nano parsley)	63.38 ± 3.22^{bc}	61.42 ± 4.54^{b}

Data were founded as mean ± SE values in the same column for each group were ($P<0.05$).

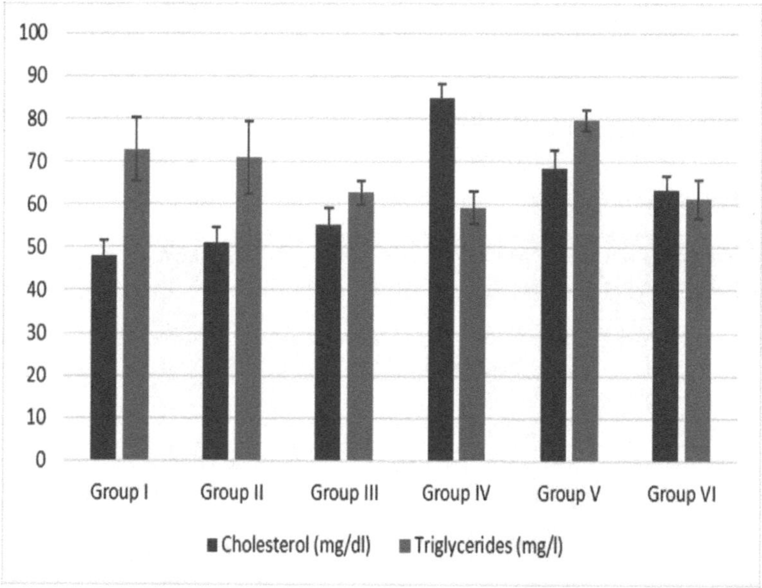

Fig. (13): Histogram showing variations in cholesterol (mg/dl), and Triglycerides (mg/l) levels of *O. niloticus* exposed to ZnO-NPs (8 mg/L) and treated with parsley leaves and parsley nanoparticles (4g/kg diet) for 6 weeks.

(High-density lipoprotein) HDL:

The results presented in Table (8) and Fig (14) showed significant ($P<0.05$) increase in HDL enzyme of group (IV). Also, exposure of Nile tilapia with ZnO-NPs (8 mg/L) and treatment with parsley leaves and parsley nanoparticles (4g/kg diet) for 6 weeks increased the level of HDL.

(Low-density lipoprotein) LDL:

LDL in serum samples of *O. niloticus* exposed to ZnO-NPs (8 mg/L) alone or co-administrated with parsley leaves and parsley nanoparticles (4g/kg diet) for 6 weeks are recorded in Table (8) and Fig (14). The results showed that the level of LDL was markedly ($P<0.05$) increased in group (IV). Meanwhile, exposure of *O. niloticus* with 8 mg/L ZnO -NPs and treated with parsley leaves and parsley nanoparticles (4g/kg diet) reduced the level of LDL to the normal level of the control groups (II and III)

(Very low-density lipoproteins) VLDL:

The mean values of VLDL in serum samples of Nile tilapia exposed to ZnO-NPs (8 mg/L) alone or co-administrated with parsley leaves and parsley nanoparticles (4g/kg diet) for 6 weeks are demonstrated in Table (8) and Fig (14). The statistical analysis showed that the level of VLDL was not changed in group (IV). And, exposed Nile tilapia with ZnO-NPs (8 mg/L) which treated with parsley leaves and parsley nanoparticles (4g/kg diet) showed also no change no change in the level of VLDL.

Table (8): Effect of ZnO-NPs (8 mg/L) on HDL, LDL, and VLDL of Nile tilapia *O. niloticus* and treated with parsley leaves and parsley nanoparticles (4g/kg diet) for 6 weeks.

Parameters / Groups	HDL (mg/dl)	LDL (mg/dl)	VLDL (mg/dl)
Group I (Control)	15.82 ± 2.18^{b}	16.8 ± 4.17^{b}	14.58 ± 1.5^{ab}
Group II (Parsley)	15.86 ± 1.68^{b}	23.06 ± 5.35^{b}	14.2 ± 1.66^{ab}
Group III (Nano parsley)	20.23 ± 1.67^{ab}	22.33 ± 2.83^{b}	12.7 ± 0.43^{ab}
Group IV (ZnO-NPs)	23.01 ± 1.07^{a}	48.83 ± 3.24^{a}	11.85 ± 0.75^{b}
Group V (ZnO-NPs +Parsley)	24.02 ± 2.08^{a}	28.58 ± 6.08^{b}	15.88 ± 0.48^{a}
Group VI (ZnO-NPs +Nano parsley)	22.45 ± 1.24^{a}	28.22 ± 3.78^{b}	12.28 ± 0.8^{b}

Data were founded as mean ± SE values in the same column for every group are ($P<0.05$).

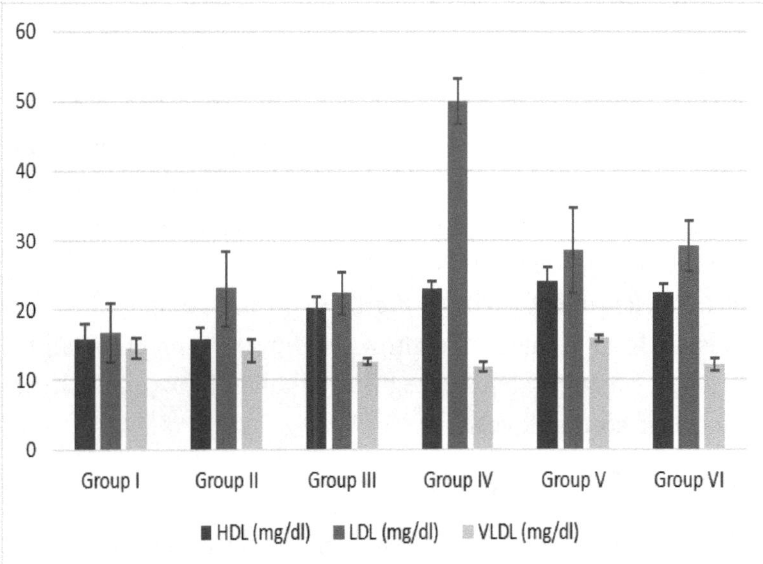

Fig. (14): Histogram showing variations in HDL (mg/dl), LDL (mg/dl) and VLDL (mg/dl) levels of *O. niloticus* exposed to ZnO-NPs (8 mg/L) and treated with parsley leaves and parsley nanoparticles (4g/kg diet) for 6 weeks.

Oxidative stress and antioxidant enzymes:

Catalase enzyme (CAT):

The mean values of liver tissue CAT in serum samples of *O. niloticus* exposed to ZnO-NPs (8 mg/L) alone or co-administrated with parsley leaves and parsley nanoparticles (4g/kg diet) for 6 weeks were demonstrated in Table (9) and Fig. (15). The results showed that the CAT enzyme levels in liver tissues of *O. niloticus* were significantly decreased ($P<0.05$) after 6 weeks of ZnO-NPs (8 mg/L) exposure compared to the tissues of liver tissues of control. However, supplementation with parsley nanoparticles upregulated the CAT enzyme in liver tissues near normal levels.

Reduced glutathione (GSH):

The mean values of liver tissue GSH in serum samples of *O. niloticus* exposed to ZnO-NPs (8 mg/L) alone or co-administrated with parsley leaves and parsley nanoparticles (4g/kg diet) for 6 weeks are demonstrated in Table (9) and fig (15). Levels of GSH were markedly decreased ($P<0.05$) in the liver tissues of group (IV) as compared to the tissues of the liver of control. So, parsley leaves and parsley nanoparticles administration improved GSH levels to nearly normal levels.

Total anti-oxidant capacity (TAC):

The mean values of liver tissue TAC in *O. niloticus* exposed to ZnO-NPs (8 mg/L) alone or co-administrated with parsley leaves and parsley nanoparticles (4g/kg diet) for 6 weeks were demonstrated in Table (9) and Fig. (16). Levels of (TAC) in liver tissues of *O. niloticus* were markedly increased ($P< 0.05$) after 6 weeks of ZnO-NPs (8 mg/L) exposure compared to the tissues of the liver control group. On the other hand, the supplementation of parsley leaves and parsley nanoparticles (4g/kg diet) enhanced the levels of TAC particularly, in group (VI) which received the parsley nanoparticles for 6 weeks.

Superoxide dismutase (SOD):

The mean values of liver tissue SOD in samples of *O. niloticus* exposed to ZnO-NPs (8 mg/L) alone or co-treated with parsley leaves and parsley nanoparticles (4g/kg diet) for 6 weeks are shown in Table (9) and Fig. (15). Levels of SOD were markedly decreased ($P<0.05$) in the liver tissues in group (IV) as compared to the control group. However, parsley leaves and parsley nanoparticles administration improved SOD to be close to normal levels.

Table (9): Effect of ZnO-NPs (8 mg/L) on oxidative stress and antioxidant enzymes in the liver of Nile tilapia, *O. niloticus* and treated with parsley leaves and parsley nanoparticles (4g/kg diet) for 6 weeks.

Parameters / Groups	CAT (Ug/mg)	GSH (nmol/mg)	TAC (Umol/mg)	SOD (Ug/mg)
Group I (Control)	22.28±2.47^{a}	33.55±2.78^{a}	0.78±0.03^{d}	17.06±0.84^{a}
Group II (Parsley)	14.12±1.11^{bc}	22.16±2.04^{ab}	1.06±0.08^{bc}	12.72±1.48^{b}
Group III (Nano parsley)	21.84±2.85^{a}	24.56±10.56^{ab}	1.1±0.02^{bc}	11.18±0.83^{bc}
Group IV (ZnO-NPs)	8.71±0.4^{c}	13.22±0.78^{b}	2.22±0.04^{a}	8.15±0.18^{c}
Group V (ZnO-NPs +Parsley)	8.63±0.38^{c}	23.56±1.44^{ab}	1.25±0.11^{b}	12.77±0.36^{b}
Group VI (ZnO-NPs +Nano parsley)	16.84±1.07^{ab}	25.68±6.33^{ab}	1.02±0.03^{c}	12.14±0.81^{bc}

Data were founded as mean ± SE values in the same column for each group were ($P<0.05$).

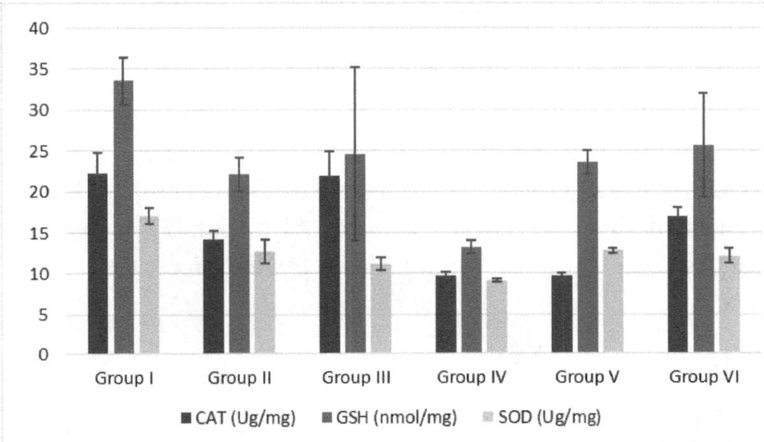

Fig. (15) Histogram showing the variations in liver GSH (nmol/mg), and CAT (Ug/mg), GSH (nmol/mg), and SOD (Ug/mg) levels of *O. niloticus* exposed to ZnO-NPs (8 mg/L) and treated with parsley leaves and parsley nanoparticles (4g/kg diet) for 6 weeks.

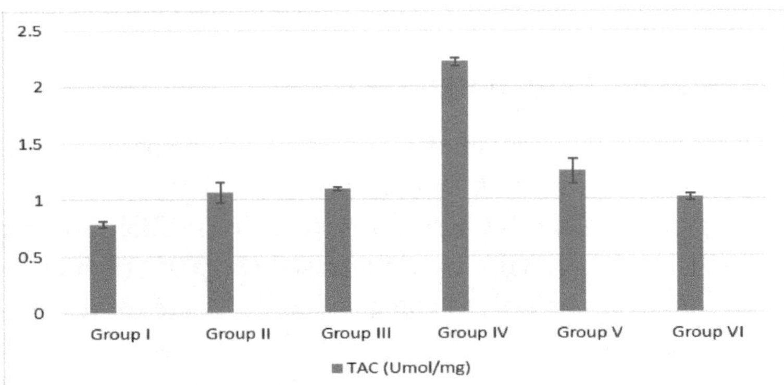

Fig. (16) Histogram showing the variations in liver TAC (Umol/mg) levels of *O. niloticus* exposed to ZnO-NPs (8 mg/L) and treated with parsley leaves and parsley nanoparticles (4g/kg diet) for 6 weeks.

Malondialdehyde (MDA):

The results presented in Table (10) and Fig. (17) showed that significantly increased MDA level increased significantly ($P<0.05$) in the liver tissues of Nile tilapia in group (IV) after exposure to ZnO -NPs (8 mg/L) for 6 weeks as compared to the tissues of the liver tissues control. Also, parsley leaves and parsley nanoparticles (4g/kg diet) treated decreased MDA levels.

Nitrous oxide (NO):

The mean values of liver tissues NO in *O. niloticus* exposed to ZnO-NPs (8 mg/L) alone or co-administrated with parsley leaves and parsley nanoparticles (4g/kg diet) for 6 weeks are demonstrated in Table (10) and fig (17). Levels of NO were markedly decreased ($P<0.05$) in the liver tissues in group (IV) as compared to the tissues of the liver of control. Meanwhile, parsley nanoparticles group (VI) increased NO levels near to levels of group (II)

Glutathione Peroxidase (GPx):

GPx levels in liver samples of Nile tilapia exposed to ZnO-NPs (8 mg/L) alone or co-treated with parsley leaves and parsley nanoparticles (4g/kg diet) are recorded in Table (10) and Fig. (17). Levels of GPx were decreased ($P< 0.05$) in the liver tissues in the group (IV) as compared to the tissues of the liver tissues control group. Whenever, parsley leaves and parsley nanoparticles (4g/kg diet) administration improved GPx levels, especially, in fish of group (VI).

Table (10): Effect of ZnO-NPs (8 mg/L) on oxidative stress and antioxidant enzymes in the liver of Nile tilapia, *O. niloticus* and treated with parsley leaves and parsley nanoparticles (4g/kg diet) for 6 weeks.

Parameters / Groups	MDA (nmol/g)	NO (µmol/ L)	GPx (mg/g)
Group I (Control)	23.28±6.56[b]	46.88±3.88[a]	88.02±3.15[a]
Group II (Parsley)	57.68±2.32[a]	30.63±1.38[b]	48.1±5.13[b]
Group III (Nano parsley)	66.38±4.88[a]	26.68±1.18[b]	80.53±1.18[a]
Group IV (ZnO-NPs)	81.63±2.38[a]	27.37±3.44[b]	30.38±2.84[c]
Group V (ZnO-NPs +Parsley)	58.12±6.41[a]	24.18±1.05[b]	58.16±2.84[b]
Group VI (ZnO-NPs +Nano parsley)	64.38±18.63[a]	31.28±1.72[b]	58.73±4.51[b]

Data were founded as mean ± SE values in the same column for each group were ($P<0.05$).

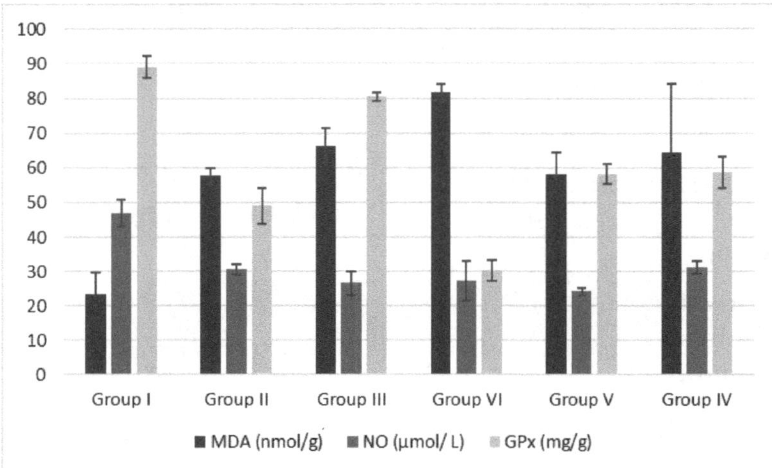

Fig. (17) Histogram showing the variations in liver MDA (nmol/g), NO (µmol/ L), and GPx (mg/g) levels of *O. niloticus* exposed to ZnO-NPs (8 mg/L) and treated with parsley leaves and parsley nanoparticles (4g/kg diet) for 6 weeks.

Immunoglobulins:

Immunoglobulin M (IgM):

The mean values of IgM in serum samples of Nile tilapia exposed to ZnO-NPs (8 mg/L) alone or co-treated with parsley leaves and parsley nanoparticles (4g/kg diet) for 6 weeks were demonstrated in Table (11) and Fig. (18). Results of IgM concentration showed marked increase after 6 weeks of ZnO-NPs exposure. Alternatively, treatment of fish with parsley leaves and parsley nanoparticles (4g/kg diet) returned IgM to nearly normal levels and this improvement was more pronounced in fish that received the parsley nanoparticles (4g/kg diet) in comparison with the fish exposed to ZnO-NPs (8 mg/L) in alone group (IV).

Immunoglobulin G (IgG):

The mean values of Immunoglobulin G (IgG) in serum samples of *O. niloticus* exposed to ZnO-NPs (8 mg/L) alone or co-administrated with parsley leaves and parsley nanoparticles (4g/kg diet) for 6 weeks were demonstrated in Table (11) and Fig. (18). Immunoglobulin G showed marked decrease after 6 weeks of ZnO-NPs (8 mg/L) exposure in comparison with the values of the control group. Meanwhile, supplementation with parsley leaves and parsley nanoparticles (4g/kg diet) increased the level of IgG near the normal levels and this enhancement was more evident in fish that received parsley nanoparticles (4g/kg diet) for 6 weeks of exposure.

Table (11): Effect of ZnO-NPs (8 mg/L) on Immunoglobulins (M+G) IgM and IgG (mg/dl) of Nile tilapia, *O. niloticus* and treated with parsley leaves and parsley nanoparticles (4g/kg diet) for 6 weeks.

Parameters / Groups	IgM(mg/dl)	IgG(mg/dl)
Group I (Control)	12.45 ± 1.83^{c}	10.87 ± 0.75^{a}
Group II (Parsley)	8.58 ± 0.28^{c}	7.87 ± 0.40^{d}
Group III (Nano parsley)	8.88 ± 0.18^{c}	8.84 ± 0.34^{ab}
Group IV (ZnO-NPs)	30.36 ± 1.18^{a}	7.4 ± 0.48^{d}
Group V (ZnO-NPs +Parsley)	10.54 ± 0.51^{c}	8.17 ± 0.28^{cd}
Group VI (ZnO-NPs +Nano parsley)	17.73 ± 2.46^{b}	8.47 ± 0.2^{bc}

Data were founded as mean ± SE values in the same column for each group were ($P<0.05$).

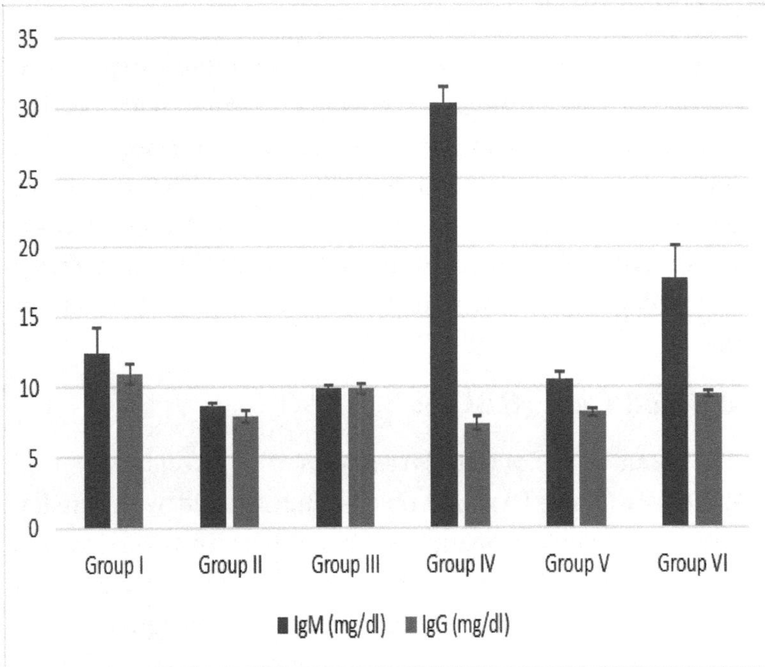

Fig. (18) Histogram showing the variations in IgM (mg/dl) and IgG (mg/dl) levels of *O. niloticus* exposed to ZnO-NPs (8 mg/L) and treated with parsley leaves and parsley nanoparticles (4g/kg diet) for 6 weeks.

Blood parameters:

Haemoglobin (Hb):

The mean values of Hb content in *O.niloticus* exposed to ZnO-NPs (8 mg/L) alone or co-supplemented with parsley leaves and parsley nanoparticles (4g/kg diet) for 6 weeks were demonstrated in Table (12) and Fig. (19). The results of revealed that Hb significantly ($P<0.05$) increased after 6 weeks. Alternatively, the treatment of fish with parsley leaves in group (V) caused a significant reduction in Hb to near the control values.

Red Blood Cells (RBCs):

The mean values of (RBCs) samples of *O. niloticus* exposed to ZnO-NPs (8 mg/L) alone or co-administrated with parsley leaves and parsley nanoparticles (4g/kg diet) for 6 weeks were demonstrated in Table (12) and Fig. (19). the statistical analysis showed that the number of (RBCs) was no changed after 6 weeks of exposure. While co-administration of Nile tilapia with parsley leaves (4g/kg diet) along with ZnO-NPs resulted in a significant decrement ($P<0.05$) in (RBCs) when compared to fish exposed to ZnO-NPs alone. But the increase in (RBCs) was clear in group (VI) which was exposed to ZnO-NPs and co-administrated with parsley nanoparticles (4g/kg diet) for 6 weeks.

White Blood Cells (WBCs) :

The mean values of WBCs of *O. niloticus* exposed to ZnO-NPs (8 mg/L) alone or co-treated with parsley leaves and parsley nanoparticles (4g/kg diet) for 6 weeks, demonstrated in Table (12) and fig. (20). The number of WBCs not change in group (IV) as compared to the control. However, parsley nanoparticles administration (4g/kg diet) increased the number of WBCs.

Platelets:

The number of Platelets in Nile tilapia exposed to ZnO-NPs (8 mg/L) alone or co-supplemented with parsley leaves and parsley nanoparticles (4g/kg diet) for 6 weeks is recorded in Table (12) and Fig. (21). The number of Platelets was decreased ($P<0.05$) in group (IV) as compared to the control group. However, parsley leaves administration increased platelets count.

Table (12): Effect of ZnO-NPs (8 mg/L) on Hb, RBCs, WBCs, and Platelets of Nile tilapia, *O. niloticus* and treated with parsley leaves and parsley nanoparticles (4g/kg diet) for 6 weeks.

Parameters / Groups	Hb (g/dl)	RBCs (Cells/mcL)	WBCs (Cells/mcL)	Platelets (Cells/mcL)
Group I (Control)	6.8 ± 0.24^{c}	3.25 ± 0.11^{b}	31.15 ± 1.1^{b}	280.84 ± 10.25^{a}
Group II (Parsley)	6.25 ± 0.21^{c}	3.03 ± 0.11^{b}	22.27 ± 4.32^{c}	177.78 ± 6.26^{c}
Group III (Nano parsley)	7.87 ± 0.28^{b}	3.0 ± 0.11^{b}	46.58 ± 1.64^{a}	238.13 ± 8.38^{b}
Group IV (ZnO-NP)	8.13 ± 0.7^{b}	3.36 ± 0.12^{b}	28.47 ± 1.04^{b}	167.02 ± 5.88^{c}
Group V (ZnO-NPs +Parsley)	6.47 ± 0.23^{c}	2.61 ± 0.08^{c}	34.48 ± 1.21^{b}	253.01 ± 8.81^{b}
Group VI (ZnO-NPs +Nano parsley)	10.03 ± 0.16^{a}	3.81 ± 0.14^{a}	43.26 ± 1.52^{a}	164.86 ± 5.81^{c}

Data were founded as mean ± SE values in the same column for each group were ($P<0.05$).

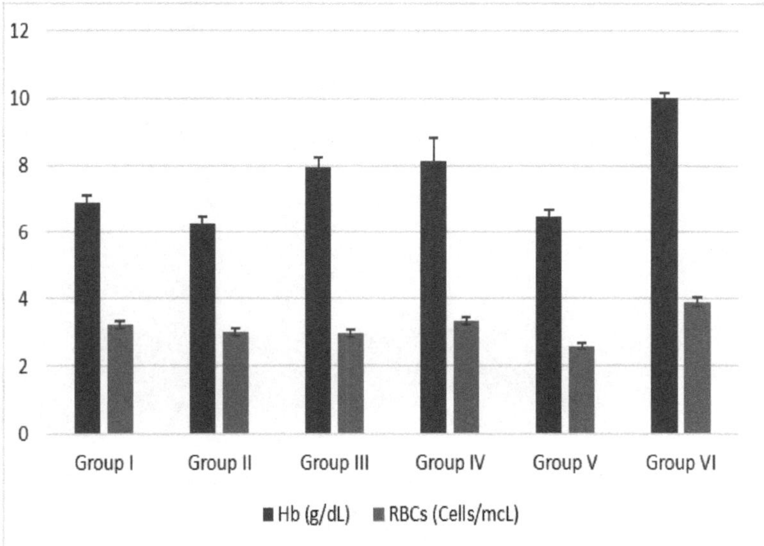

Fig. (19) Histogram showing the variations in Hb (g/dL), the number of RBCs (cells/mcL) *O. niloticus* exposed to ZnO-NPs (8 mg/L) and treated with parsley leaves and parsley nanoparticles (4g/kg diet) for 6 weeks.

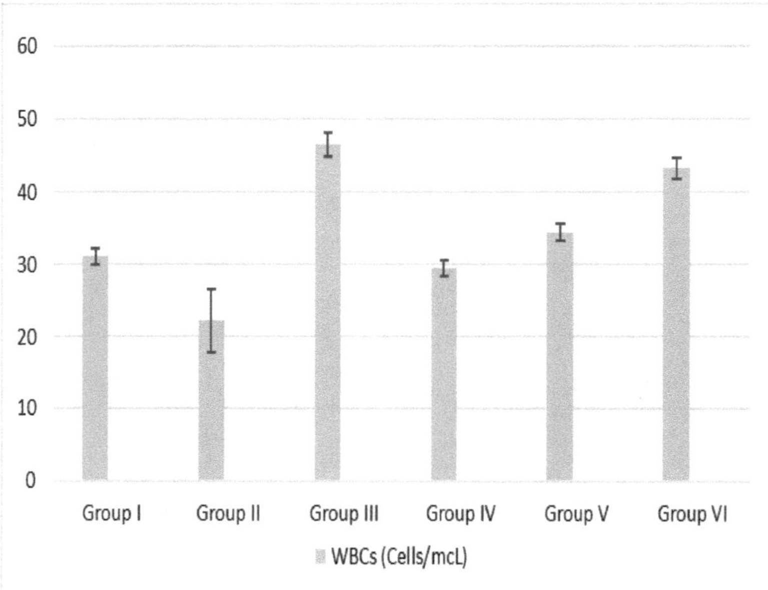

Fig. (20) Histogram showing the variations in the number of WBCs (cells/mcL) *O. niloticus* exposed to ZnO-NPs (8 mg/L) and treated with parsley leaves and parsley nanoparticles (4g/kg diet) for 6 weeks.

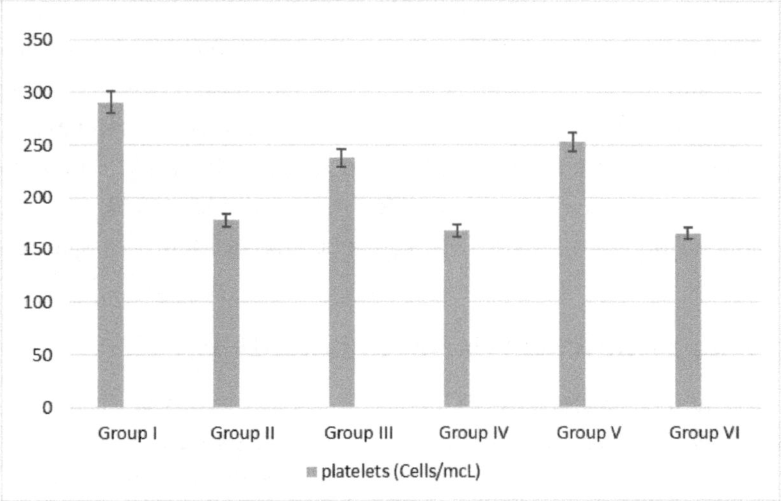

Fig. (21) Histogram showing the variations in Platelets count (cells/mcL) *O. niloticus* exposed to ZnO-NPs (8 mg/L) and treated with parsley leaves and parsley nanoparticles (4g/kg diet) for 6 weeks.

Mean corpuscular volume (MCV):

The mean values of MCV in Nile tilapia exposed to (8 mg/L) (ZnO-NPs) alone or co-treated with parsley leaves and parsley nanoparticles (4g/kg diet) for 6 weeks are demonstrated in Table (13) and Fig. (22). The results revealed that the MCV was insignificantly changed ($P<0.05$) in group (IV). However, exposure of Nile tilapia with 8 mg/L (ZnO-NPs) and treatment with parsley leaves and parsley nanoparticles (4g/kg diet) reduced mean values of MCV, especially the fish exposed to parsley leaves (4g/kg diet) (group V).

Mean corpuscular Haemoglobin (MCH):

The mean values of MCH in *O. niloticus* exposed ZnO-NPs (8 mg/L) alone or co-administrated with parsley leaves and parsley nanoparticles (4g/kg diet) for 6 weeks are presented in Table (13) and Fig. (22). The results of showed that (MCH) was significantly ($P<0.05$) decreased after 6 weeks of *O. niloticus* exposure to ZnO-NPs. Alternatively, the treatment of fish with parsley leaves group (V) and parsley nanoparticles group (VI) showed a significant increment in (MCH) values. This was more evident in group (VI) which received parsley nanoparticles (4g/kg diet) compared with the control group

Mean Corpuscular Haemoglobin Concentration (MCHC):

MCHC values of Nile tilapia exposed to ZnO-NPs (8 mg/L) alone or co-treated with parsley leaves and parsley nanoparticles (4g/kg diet) for 6 weeks are demonstrated in Table (13) and Fig (22). The statistical analysis recorded that the mean values of MCHC were markedly ($P<0.05$) decreased in group (IV). however, exposure of Nile tilapia to ZnO-NPs (8 mg/L) and treatment with parsley leaves and parsley nanoparticles (4g/kg diet) increased the values of MCHC. The results were more obvious in group (VI).

Table (13): Effect of ZnO-NPs (8 mg/L) on MCV, MCH, and MCHC of Nile tilapia, *O. niloticus* and treated with parsley leaves and parsley nanoparticles (4g/kg diet) for 6 weeks.

Parameters / Groups	MCV (fL)	MCH (pg/cell)	MCHC (g/dI)
Group I (Control)	116.58±4.1[a]	27.81±0.88[cd]	18.32±0.65[e]
Group II (Parsley)	85.8±3.71[cd]	26.38±2.2[d]	28.08±1.02[c]
Group III (Nano parsley)	106.67±3.76[abc]	33.18±1.17[b]	32.08±1.13[bc]
Group IV (ZnO-NPs)	111.12±3.82[ab]	20.78±0.73[e]	22.63±0.76[d]
GroupV (ZnO-NPs +Parsley)	82.34±3.25[d]	31.25±1.1[bc]	32.44±1.14[b]
Group VI (ZnO-NPs+Nano parsley)	103.54±3.64[bcd]	40.51±1.42[a]	42.02±1.48[a]

Data were founded as mean ± SE values in the same column for each group were (P<0.05).

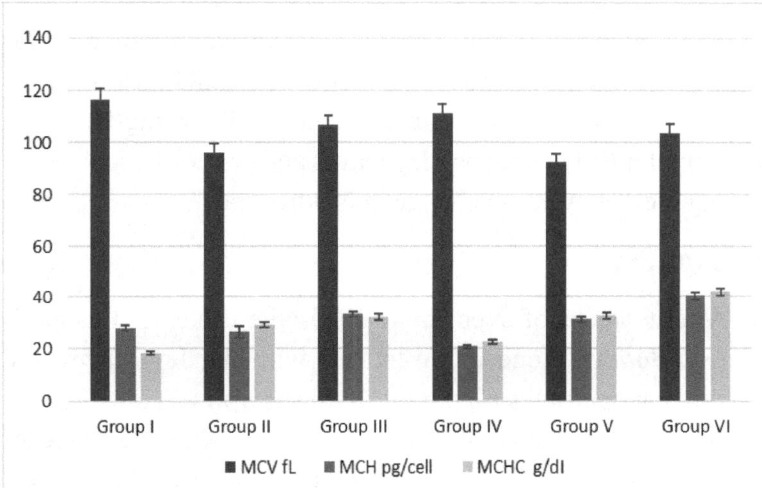

Fig. (22) Histogram showing the variations in MCV (fL), MCH (pg/cell), and MCHC (g/dl) levels of *O. niloticus* exposed to ZnO-NPs (8 mg/L) and treated with parsley leaves and parsley nanoparticles (4g/kg diet) for 6 weeks.

Neutro. staff:

From Table (14) and Fig. (23), mean value Neutro. staff showed no change after 6 weeks exposure of ZnO-NPs in group (IV). Also, exposed Nile tilapia to ZnO-NPs (8 mg/L) which was supplemented with parsley leaves and parsley nanoparticles (4g/kg diet) showed no change in Neutro. staff.

Basophils:

After 6 weeks of exposure to ZnO-NPs values of Basophils in *O. niloticus* alone or co-treated with parsley leaves and parsley nanoparticles (4g/kg diet) are shown in Table (14) and Fig. (23). The statistical analysis recorded that the value of basophils were no changed in group (IV). Meanwhile, exposed *O. niloticus* to ZnO-NPs (8 mg/L) which treated with parsley leaves and parsley nanoparticles (4g/kg diet) showed insignificant change ($p<0.05$) in the mean values of basophils copared with the control group.

Eosinophils:

Eosinophils of Nile tilapia exposed to ZnO-NPs (8 mg/L) alone or co-administrated with parsley leaves and parsley nanoparticles (4g/kg diet) for 6 weeks are demonstrated in Table (14) and Fig. (23). The statistical analysis recorded that the mean values of eosinophils were markedly ($P<0.05$) insignificant changed in group (IV). However, exposure of Nile tilapia to ZnO-NPs (8 mg/L) and co-treated with parsley nanoparticles (4g/kg diet) increased eosinophils to be close values in group (VI).

Monocytes:

It can be seen from Table (14) and Fig. (23) showed that the mean values of monocytes in Nile tilapia exposed to ZnO-NPs (8 mg/L) alone or co-treated with parsley leaves and parsley nanoparticles (4g/kg diet) for 6 weeks were markedly ($P<0.05$) decreased in the group (IV) as compared to the control group. Whenever, exposed Nile tilapia with ZnO-NPs (8 mg/L) and treatment with parsley leaves increased the values of monocytes to the normal values of the control group.

Table (14): Effect of ZnO-NPs (8 mg/L) on Neutro.staff, Basophils, Eosinophils, and Monocyte of Nile tilapia, *O. niloticus* and treated with parsley leaves and parsley nanoparticles (4g/kg diet) for 6 weeks.

Parameters / Groups	Neutro.staff	Basophils	Eosinophils	Monocytes
Group I (Control)	$2.5{\pm}0.28^{a}$	$1.5{\pm}0.28^{ab}$	$4.5{\pm}0.28^{c}$	$5.75{\pm}0.25^{a}$
Group II (Parsley)	$2.5{\pm}0.28^{a}$	$0.5{\pm}0.28^{b}$	$4.25{\pm}0.48^{c}$	$4.75{\pm}0.25^{b}$
Group III (Nano parsley)	$2.75{\pm}0.25^{a}$	$2.25{\pm}0.48^{a}$	$6{\pm}0.41^{ab}$	$4.75{\pm}0.25^{b}$
Group IV (ZnO-NP)	$3.25{\pm}0.48^{a}$	$1.25{\pm}0.25^{ab}$	$5.5{\pm}0.28^{abc}$	$3.25{\pm}0.25^{c}$
Group V (ZnO-NPs +Parsley)	$2.5{\pm}0.28^{a}$	$1.5{\pm}0.28^{ab}$	$5{\pm}0.71^{bc}$	$5.25{\pm}0.25^{ab}$
Group VI (ZnO-NPs +Nano parsley)	$2.25^{a}{\pm}0.25$	$2.25{\pm}0.48^{a}$	$6.5{\pm}0.28^{a}$	$3.25{\pm}0.25^{c}$

Data were founded as mean ± SE values in the same column for each group were (P<0.05).

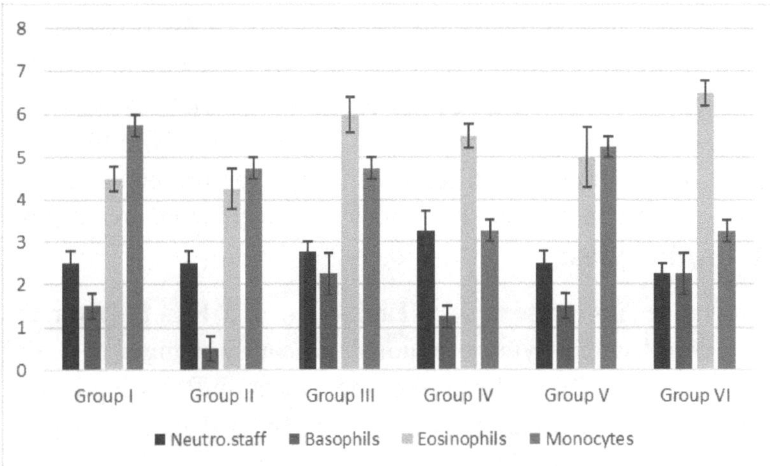

Fig. (23) Histogram showing the variations in Neutro. staff, Basophils, Eosinophils, and Monocytes levels of *O.niloticus* exposed to ZnO-NPs (8 mg/L) and treated with parsley leaves and parsley nanoparticles (4g/kg diet) for 6 weeks.

Segmented neutrophils:

Values of Segmented neutrophils of Nile tilapia exposed to (ZnO-NPs) (8 mg/L) alone or co-supplmented with parsley leaves and parsley nanoparticles (4g/kg diet) for 6 weeks are demonstrated in Table (15) and Fig. (24). The results revealed that Segmented neutrophils were not significantly changed ($P<0.05$) in group (IV). However, exposure of Nile tilapia to ZnO-NPs (8 mg/L) and treatment with parsley nanoparticles (4g/kg diet) increased the values of Segmented neutrophils

Lymphocytes:

The results presented in Table (15) and Fig. (24) for the values of Lymphocytes were not significantly changed ($P<0.05$) in group (IV). Nile tilapia exposed to ZnO-NPs (8 mg/L) and co-administrated with parsley leaves (4g/kg diet) not changed the number of Lymphocytes.

Table (15): Effect of ZnO-NPs (8 mg/L) on Segmented neutrophils and lymphocytes of Nile tilapia, *O. niloticus* and treated with parsley leaves and parsley nanoparticles (4g/kg diet) for 6 weeks.

Groups \ Parameters	Segmented neutrophils	Lymphocytes
Group I (Control)	55.75 ± 1.44^{bc}	28.75 ± 0.85^{a}
Group II (Parsley)	53.25 ± 1.7^{c}	32.75 ± 2.06^{a}
Group III (Nano parsley)	66 ± 0.81^{a}	18.5 ± 0.65^{b}
Group IV (ZnO-NP)	58.5 ± 1.32^{b}	28.25 ± 1.03^{a}
Group V (ZnO-NPs+Parsley)	56.75 ± 2.02^{bc}	28 ± 2.35^{a}
Group VI (ZnO-NPs+Nano parsley)	66 ± 1.68^{a}	18.75 ± 1.44^{b}

Data were founded as mean ± SE values in the same column for each group were ($P<0.05$).

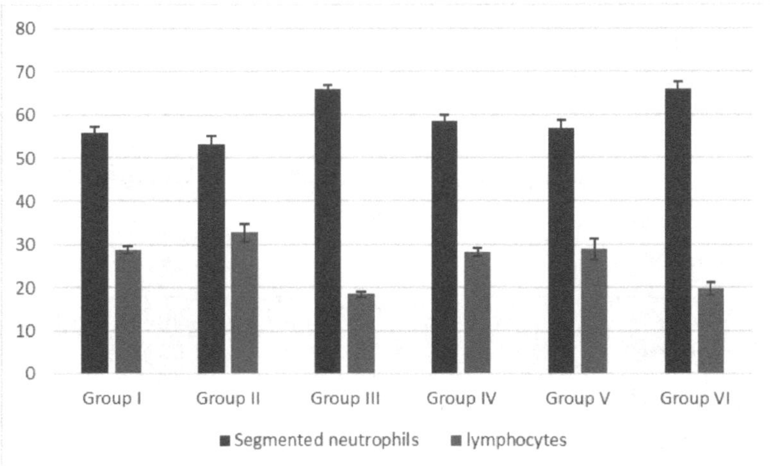

Fig. (24) Histogram showing the variations in the values of Segmented neutrophils and lymphocytes of *O. niloticus* exposed to ZnO-NPs (8 mg/L) and treated with parsley leaves and parsley nanoparticles (4g/kg diet) for 6 weeks.

Histopathological findings:

The microscopic examination of the liver of tilapia in the control group revealed polyhedral hepatocytes with clearly visible cell membranes radially arranged around the central vein. These hepatocytes contained a large spherical centrally located nucleus with vacuolar structures and the hepatic sinusoids were lined by Kupffer cells (Fig. 25). The exocrine pancreas was formed from clusters of pancreatic acini scattered throughout the hepatic parenchyma around the portal veins forming the hepatopancreas. These acini were formed from typical acinar polyhedral cells with the central narrow lumen. The nucleus was spherical and basally located and the basal cytoplasm was basophilic, while the apical cytoplasm contained eosinophilic zymogen granules (Fig. 26).

The microscopic findings of the examined liver of tilapia given parsley revealed a nearly similar microscopic picture of the liver of fish in the control group in addition to congestion of the central veins and sinusoids (Fig.27)

Also, the examined liver of tilapia given parsley showed the same microscopic findings of the liver of tilapia given parsley nanoparticles where the center veins were congested and clusters of pancreatic acini were arranged around congested portal veins (Fig. 28 & 29)

Meanwhile, the microscopic examination of the liver of tilapia intoxicated with ZnO-NPs revealed severe histopathological alterations. Severe congestion of hepatic sinusoids and pyknosis of the nuclei of the hepatocytes together with apoptosis of some hepatic cells were prevalent in most of the examined liver (Fig. 30). Multiple mononuclear inflammatory cellular infiltration mainly lymphocytes which replaced degenerated hepatic cells were seen (Fig. 31 & 32). Marked activation of Kupffer cells was observed. Recent thrombosis attached to damaged blood vessel walls and rupture of many hepatocytes forming fatty cysts were recorded in a few of the examined livers (Fig. 33). Dilatation of the blood vessels with necrosis of hepatic cells particularly around these vessels was quite evident in the liver of fish exposed to ZnO-NPs (Fig. 34 & 35). These hepatic cells showed eosinophilic debris in the cytoplasm with pyknosis or complete absence of their nuclei. Focal areas of hemorrhages replaced the hepatic parenchyma were detected (Fig. 36). In addition, bile ductal hyperplasia and periductal inflammatory cellular infiltration were found (Fig. 37).

Moreover, the examined hepatopancreatic tissues revealed severe congestion of the portal vessels with dissociation and necrosis of the pancreatic acini (Fig. 38). In few cases, focal areas of hemorrhages were recorded around the dissociated and necrotic pancreatic acini (Fig. 39). Hemolysis of erythrocytes within the portal veins was also seen (Fig. 40).

The histopathological examination of liver of tilapia intoxicated with ZnO-NPs and treated with parsley revealed improvement in the microscopic picture of the liver comparing with the examined liver of tilapia intoxicated with ZnO-NPs alone. The examined liver of this group showed multiple areas of mononuclear inflammatory cells and activation of kupffer cell (Fig. 41). Dilatation of the central veins and necrosis of few hepatocytes were prevalent (Fig. 42). In addition, some central veins were engorged with intact and hemolyzed erythrocytes (Fig. 43). The examined hepatopancreas revealed congestion of portal vessel and periportal necrosis of few hepatocytes evidenced by pyknosis and absence of their nuclei (Fig. 44). Dissociation of the pancreatic acini was recorded in some examined hepatopancreas (Fig. 45).

Also, the histopathological examination of liver of tilapia intoxicated with ZnO-NPs and treated with parsley nanoparticles revealed enhancement in the microscopic picture of the liver comparing with the examined liver of tilapia intoxicated with ZnO-NPs alone. The examined liver of this group showed congestion of central veins and sinusoids with pyknosis of nuclei of some hepatocytes (Fig. 46). Focal aggregation of few mononuclear inflammatory cells particularly lymphocytes cells were also detected in-between the hepatic (Fig. 47). Necrosis of some hepatocytes represented by the presence of eosinophilic debris in the cytoplasm (Fig. 48) and absence of nuclei of hepatocytes (Fig. 49), Moreover, congestion of portal vein and dissociation of some pancreatic acini were also seen in few examined hepato-pancreas (Fig. 50)

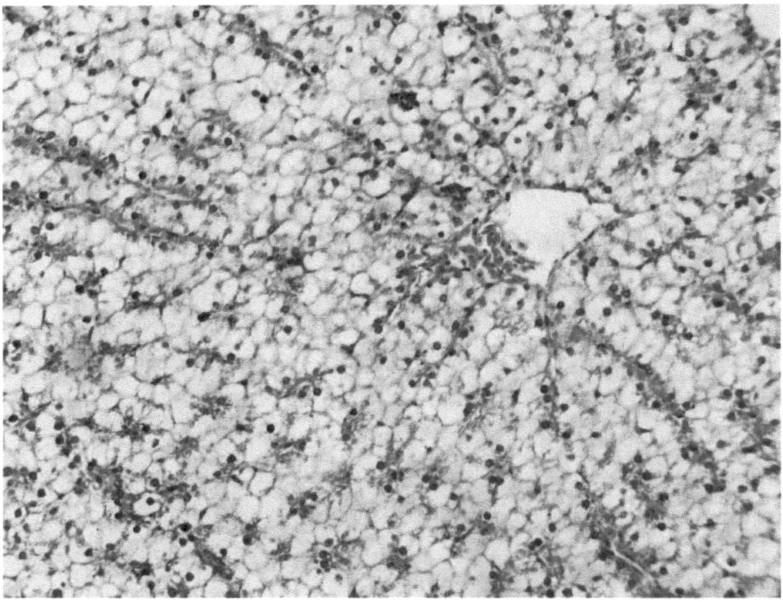

Fig. (25) Photomicrograph of the liver of Nile tilapia in control group showing polyhedral hepatocytes contained a large spherical centrally located nucleus with vacuolar structures and radially arranged around the central vein. H & E stain X200

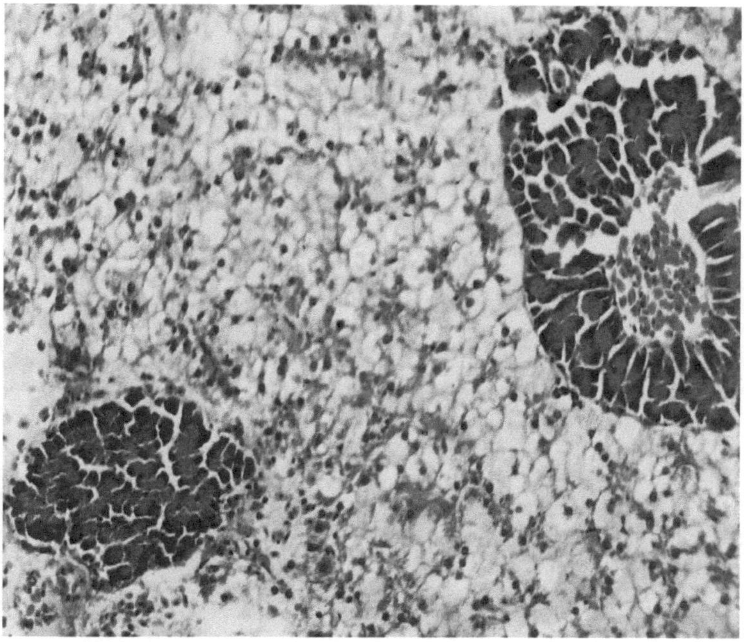

Fig. (26) Photomicrograph of the liver of Nile tilapia in control group showing clusters of pancreatic acini around the portal vein forming the hepatopancreas. H & E stain X200

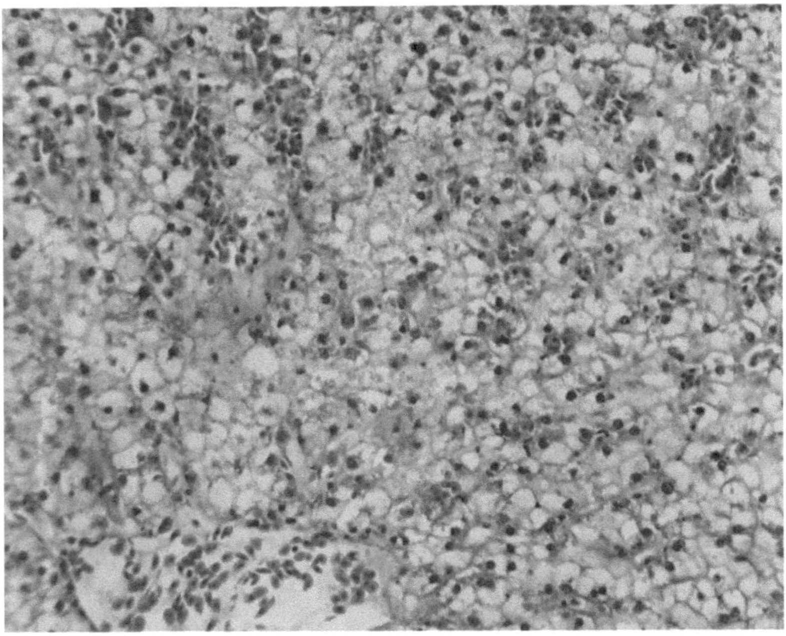

Fig. (27) Photomicrograph of the liver of Nile tilapia given parsley alone showing congestion of central vein and hepatic sinusoids H & E stain X200

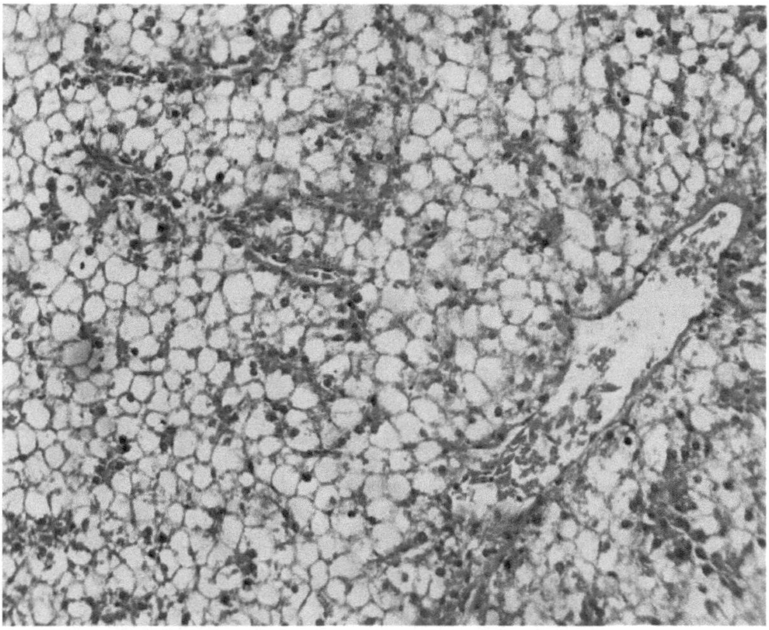

Fig. (28) Photomicrograph of the liver of Nile tilapia given parsley nanoparticle alone showing congestion of central vein H & E stain X200

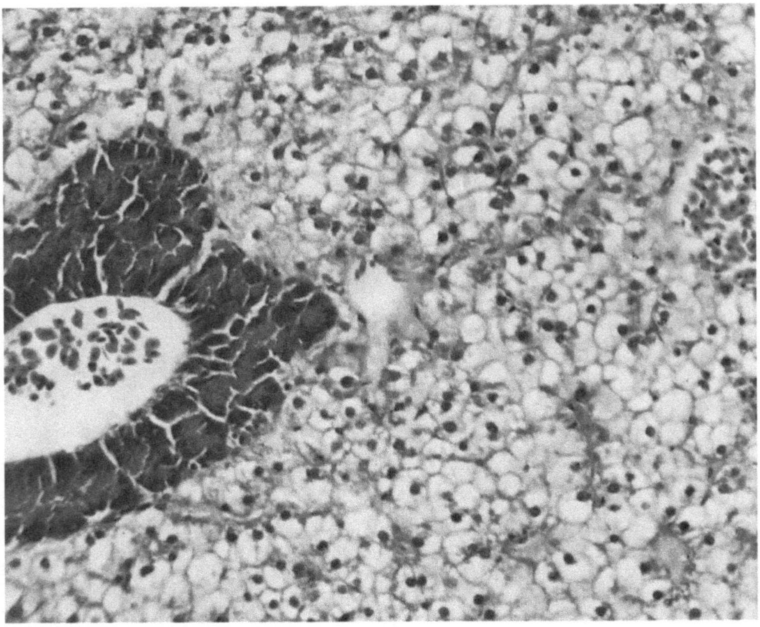

Fig. (29) Photomicrograph of the hepatopancreas of Nile tilapia given parsley nanoparticle alone showing cluster of pancreatic acini around congested portal vein. H & E stain X200

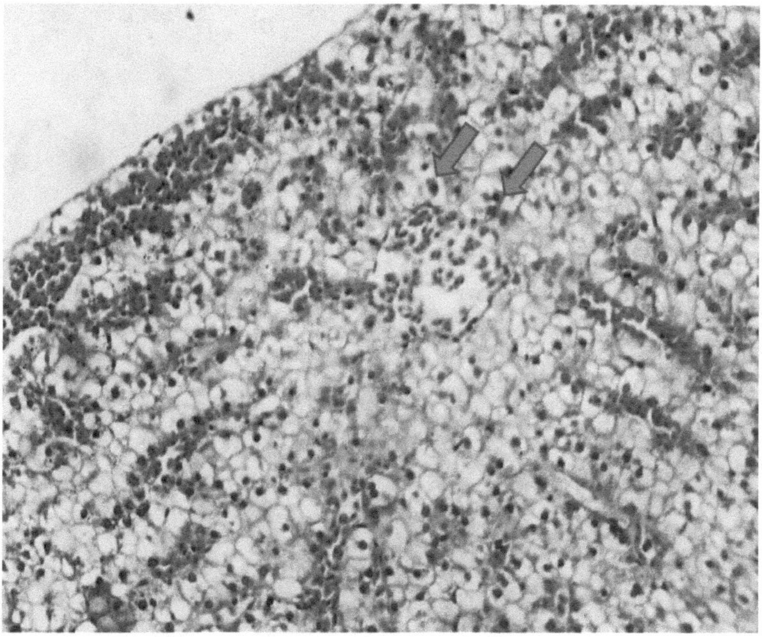

Fig. (30) Photomicrograph of the liver of Nile tilapia intoxicated with ZnO-NPs showing congestion of sinusoids pyknosis of nuclei of some hepatic cells and apoptosis of another few hepatic cells (arrow) H & E stain X200

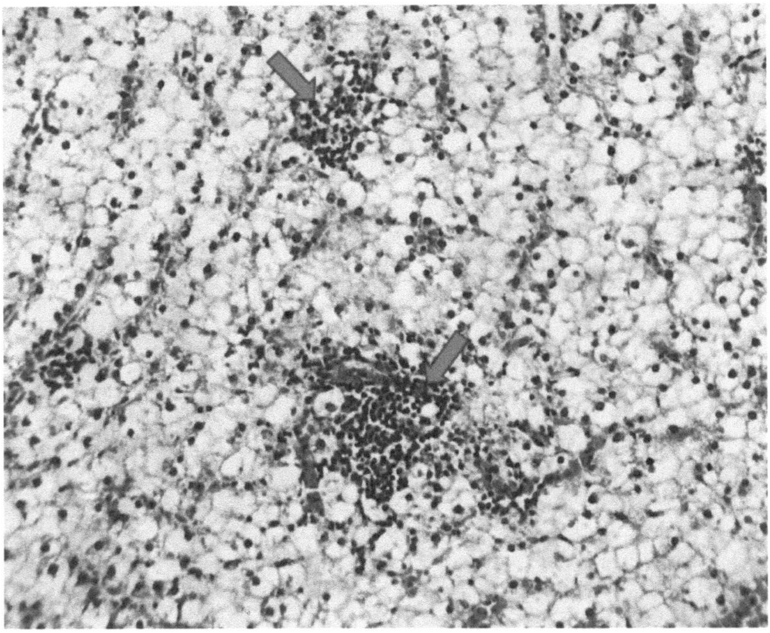

Fig. (31) Photomicrograph of the liver of Nile tilapia intoxicated with ZnO-NPs showing multiple areas of inflammatory cellular infiltration (arrow) H & E stain X200

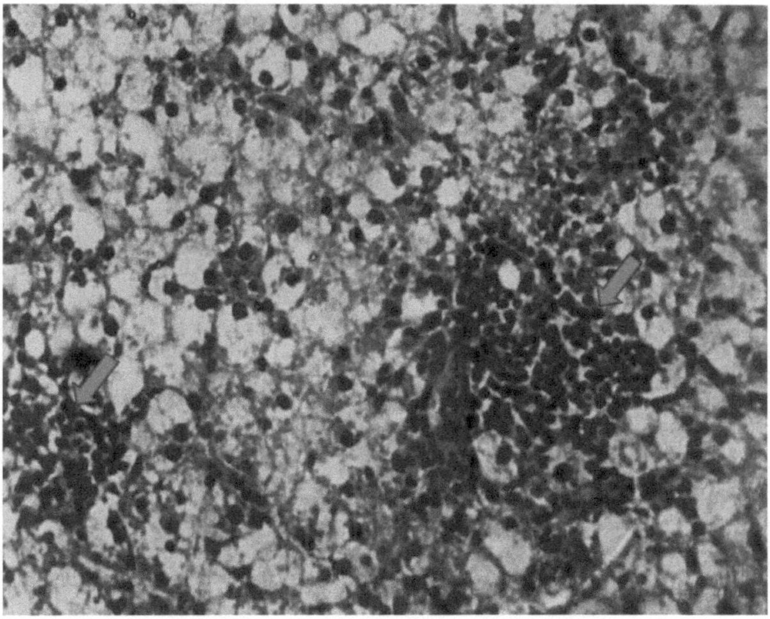

Fig. (32) High Power of the previous figure showing inflammatory cellular infiltration mainly lymphocytes replaced degenerated hepatic cells. Notice pyknosis of the nuclei of hepatic cells (arrow). H & E stain 400

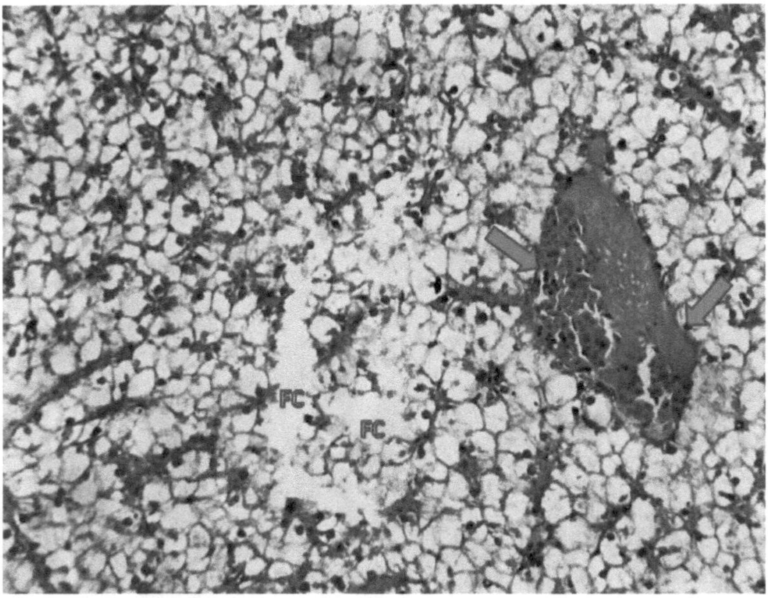

Fig. (33) Photomicrograph of the liver of Nile tilapia intoxicated with ZnO-NPs showing thrombotic mass attached to damaged blood vessel wall (arrow) with rupture of many hepatocytes forming fatty cysts (FC) H & E stain X200

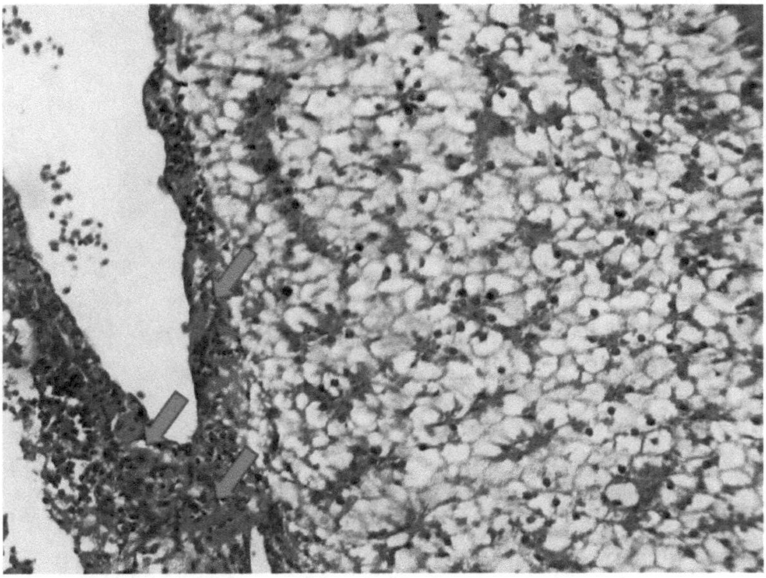

Fig. (34) Photomicrograph of the liver of Nile tilapia intoxicated with ZnO-NPs showing dilatation of the blood vessel with perivascular hepatic necrosis (arrow) H & E stain X200

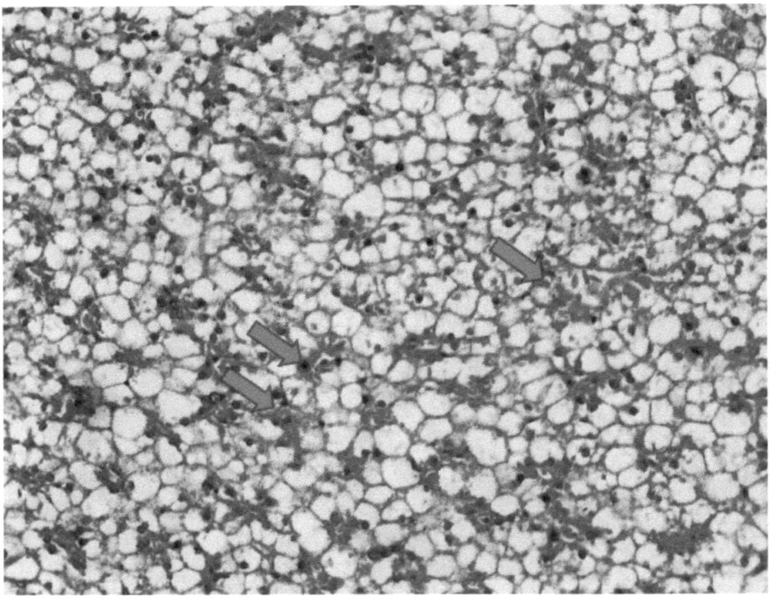

Fig. (35) Photomicrograph of the liver of Nile tilapia intoxicated with ZnO-NPs showing necrosis of hepatocytes (arrow) H & E stain X200

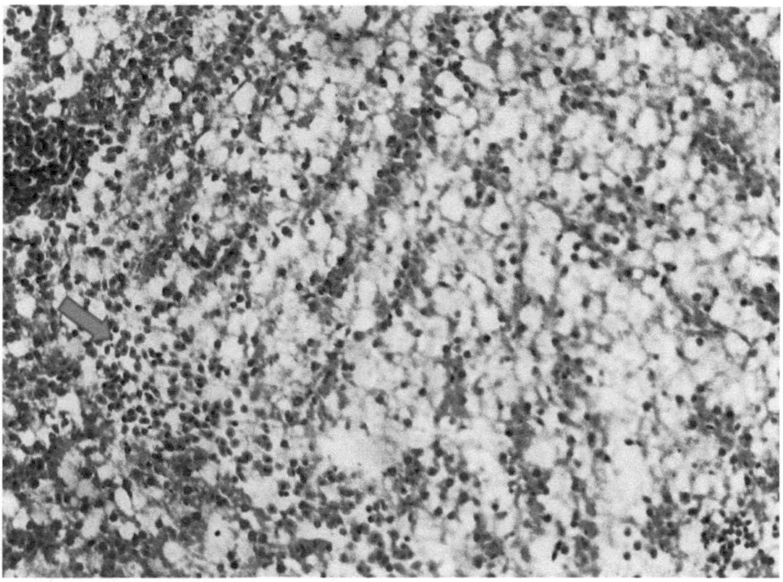

Fig. (36) Photomicrograph of the liver of Nile tilapia intoxicated with ZnO-NPs showing congestion of sinusoids with focal area of hemorrhage replaced the hepatic parenchyma (arrow) H & E stain X200.

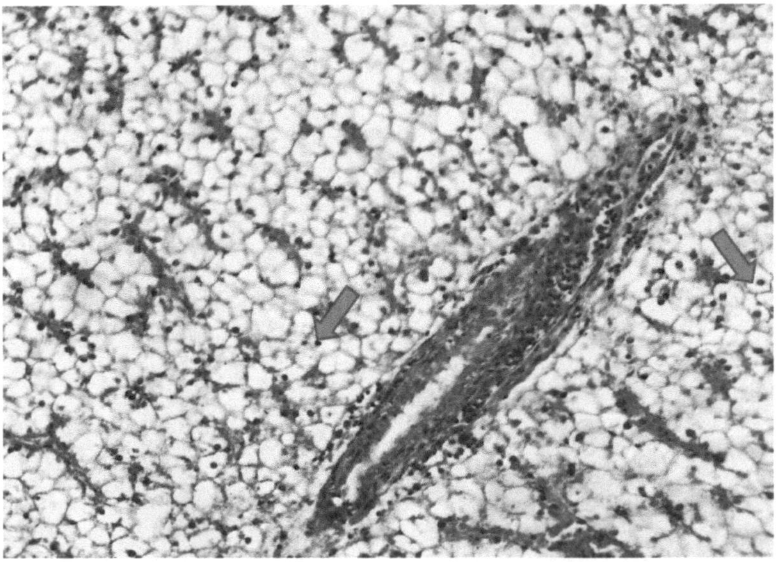

Fig. (37) Photomicrograph of the liver of Nile tilapia intoxicated with ZnO-NPs showing pyknosis of nuclei of the hepatocytes (arrow) with bile ductal hyperplasia and periductal inflammatory cellular infiltration H & E stain X200

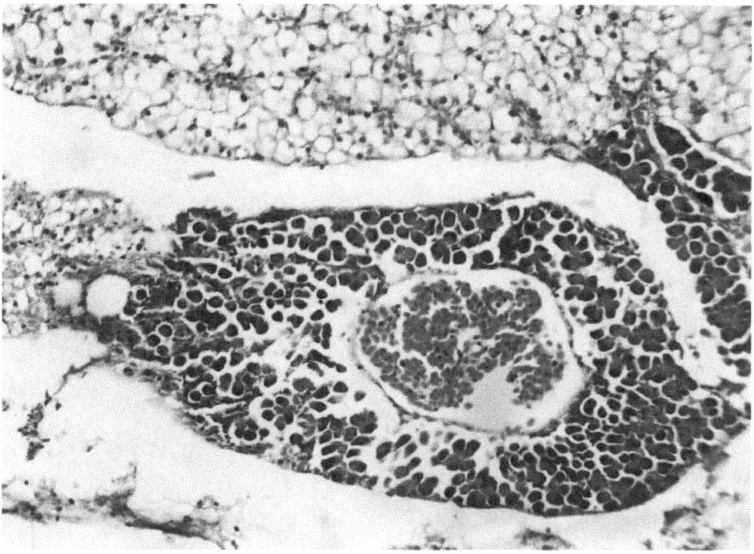

Fig. (38) Photomicrograph of the hepatopancreatic tissues of Nile tilapia intoxicated with ZnO-NPs showing congestion with dissociation and necrosis of the pancreatic acini (arrow).H & E stain X200

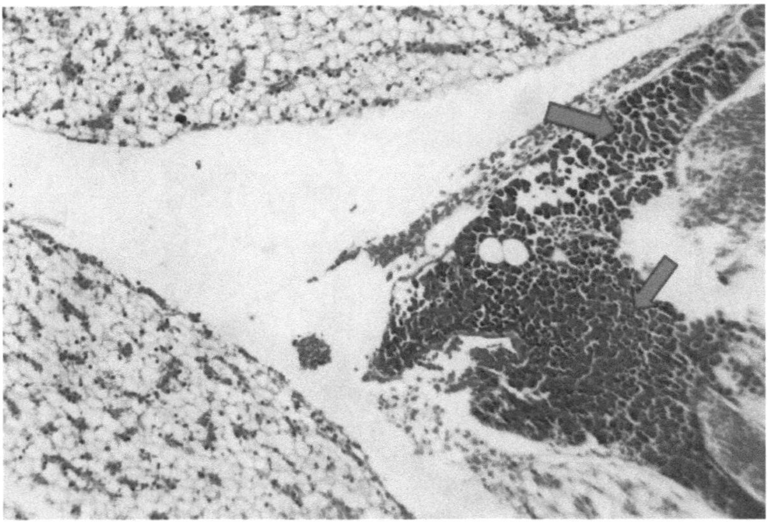

Fig. (39) Photomicrograph of the hepatopancreatic tissues of Nile tilapia intoxicated with ZnO-NPs showing congestion and hemorrhage with dissociation and necrosis of the pancreatic acini (arrow).H & E stain X200

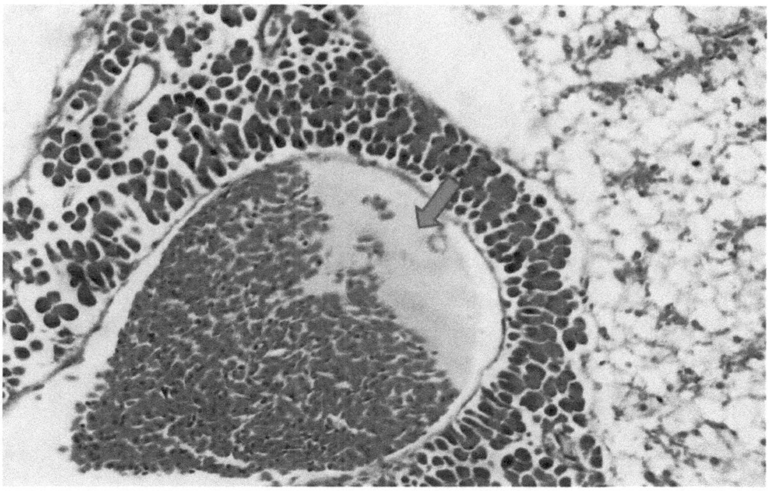

Fig. (40) Photomicrograph of the hepatopancreatic tissues of Nile tilapia intoxicated with ZnO-NPs showing congestion with hemolysis of erythrocytes (arrow) with dissociation of the pancreatic acini.H & E stain X200

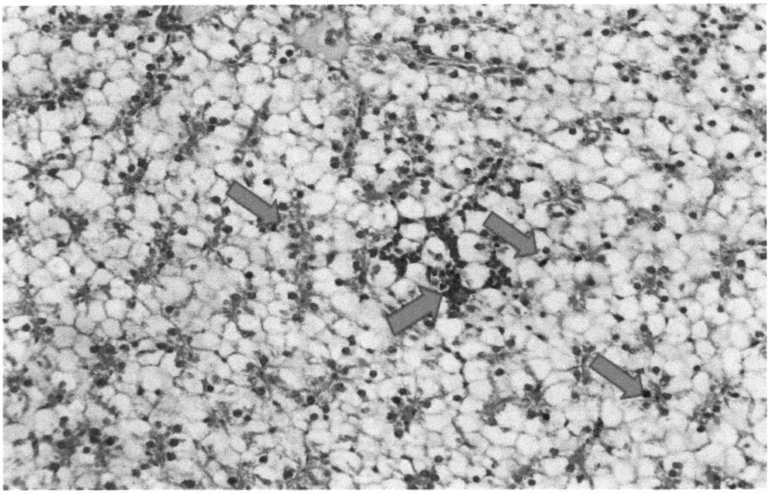

Fig. (41) Photomicrograph of the liver of Nile tilapia intoxicated with ZnO-NPs and treated with parsley showing focal aggregation of mononuclear inflammatory cells with activation of kupffer cells and necrosis of some hepatocytes (arrow) H & E stain X200.

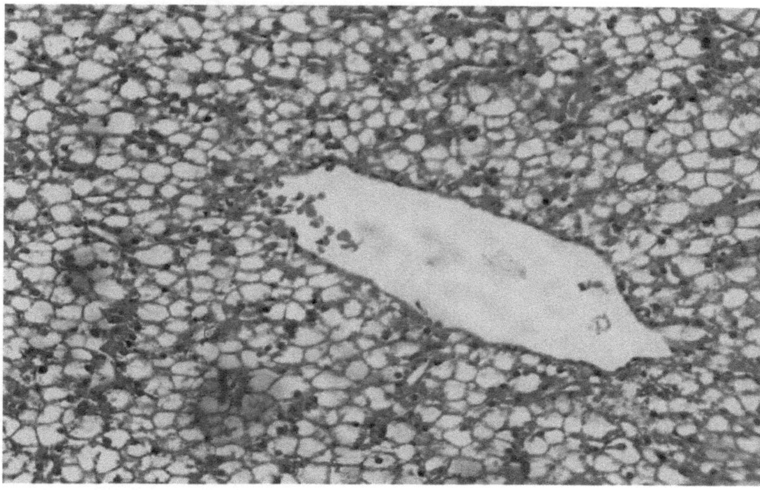

Fig. (42) Photomicrograph of the hepatopancreas of Nile tilapia intoxicated with ZnO-NPs and treated with parsley showing dilatation of the central vein and necrosis of few hepatocytes (arrow) H & E stain X200.

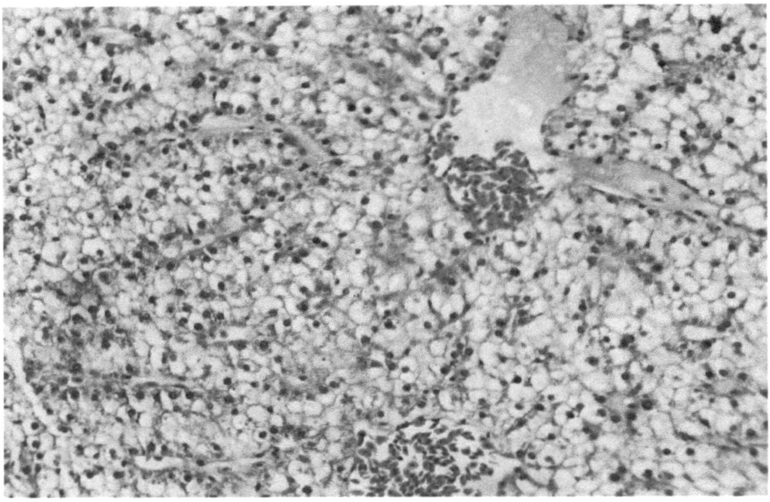

Fig. (43) Photomicrograph of the liver of Nile tilapia intoxicated with ZnO-NPs and treated with parsley showing engorgement of central veins with intact and hemolyzed erythrocytes with pyknosis of the nucle of some hepatocytes (arrow) H & E stain X200.

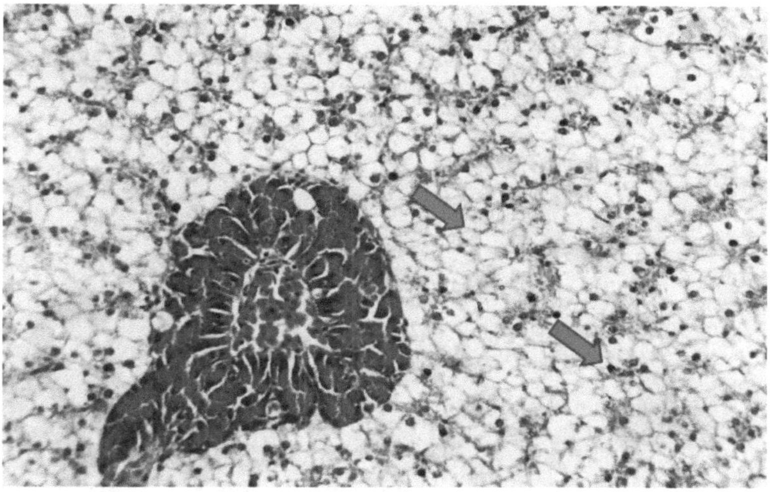

Fig. (44) Photomicrograph of the hepatopancreas of Nile tilapia intoxicated with ZnO-NPs and treated with parsley showing congestion of portal vessel with pyknosis and absence of nuclei of hepatocytes (arrow) H & E stain X200.

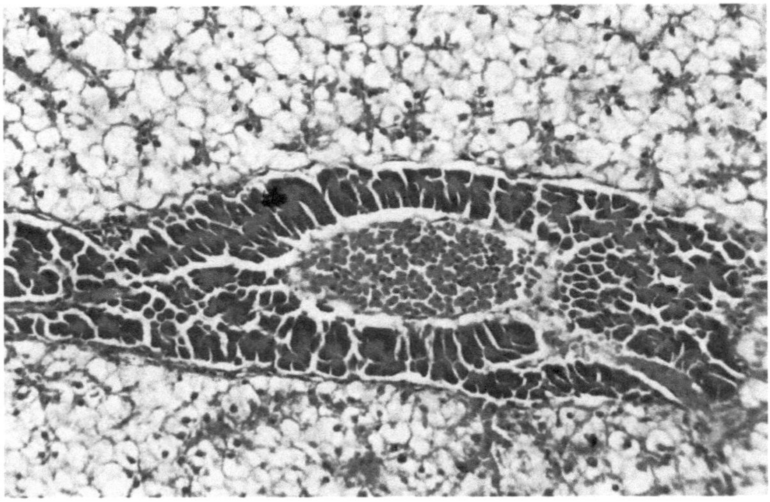

Fig. (45) Photomicrograph of the hepatopancreas of Nile tilapia intoxicated with ZnO-NPs and treated with parsley showing congestion of portal vessel and dissociation of the pancreatic acini with absence of nuclei of some hepatic cells. H & E stain X200.

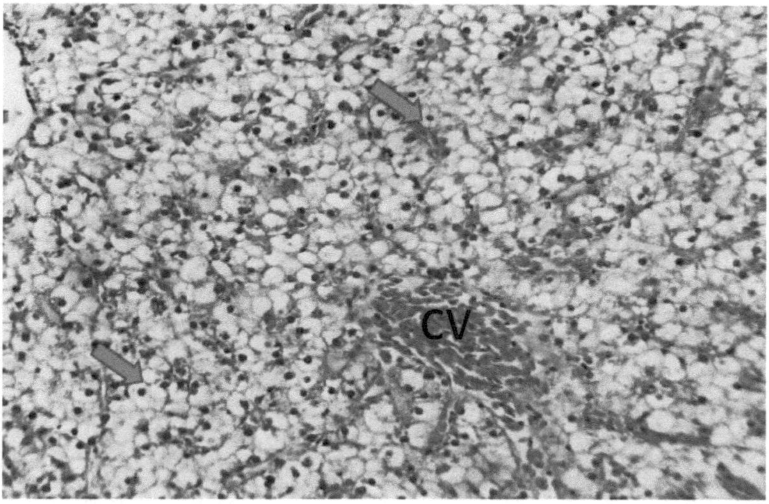

Fig. (46) Photomicrograph of the liver of Nile tilapia intoxicated with ZnO-NPs and treated with parsley nanoparticles showing congestion of central vein (CV) and sinusoids with pyknosis of nuclei of some hepatocytes (arrow) H & E stain X200

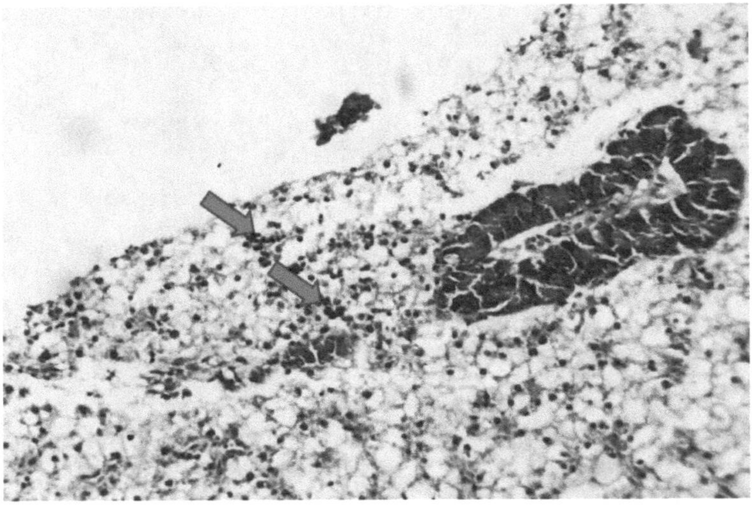

Fig. (47) Photomicrograph of the hepatopancreas of Nile tilapia intoxicated with ZnO-NPs and treated with parsley nanoparticles showing congestion of portal blood vessel with aggregation of few mononuclear inflammatory cells (arrow) H & E stain X200.

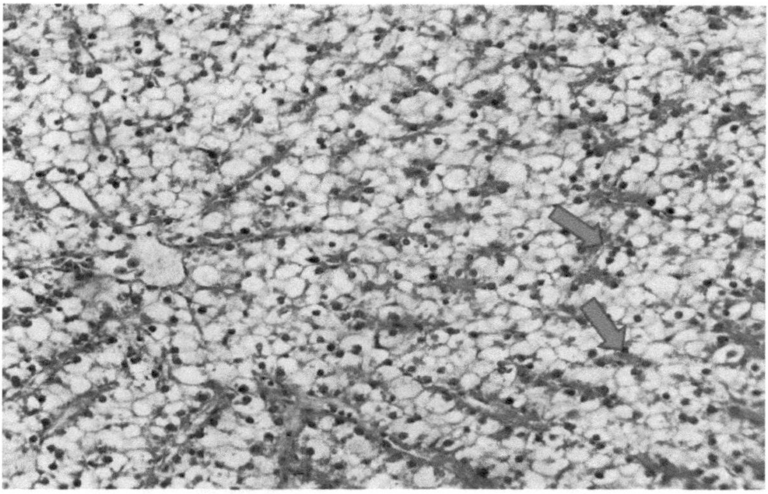

Fig. (48) Photomicrograph of the liver of Nile tilapia intoxicated with ZnO-NPs and treated with parsley nanoparticles showing eosinophilic debris in the cytoplasm of necrotic hepatocytes (arrow) H & E stain X200.

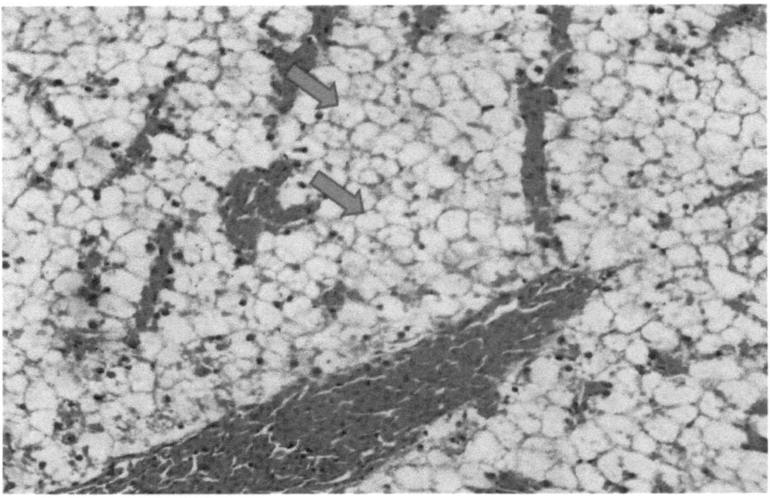

Fig. (49) Photomicrograph of the liver of Nile tilapia intoxicated with ZnO-NPs and treated with parsley nanoparticles showing congestion of blood vessels and sinusoids with absence of nuclei of hepatocytes (arrow) H & E stain X200.

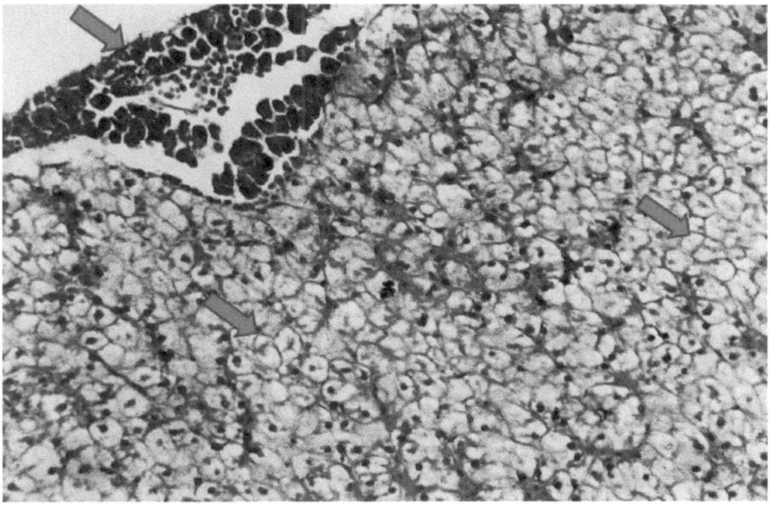

Fig. (50) Photomicrograph of the hepatopancreas of Nile tilapia intoxicated with ZnO-NPs and treated with parsley showing congestion of portal vein and dissociation of the pancreatic acini and absence of nuclei of some hepatic cells (arrow) H & E stain X200.

Discussion

DISCUSSION

Pollution is one of the environmental problems that it constitutes a major threat to humans and global ecosystems **(Brand 2001)**. Fish play a major role in the aquatic food chain as they have the ability to metabolize and accumulate minerals. **(Luoma and Rainbow 2008)**

Zinc oxide is one of the oxides that have unique properties as a physical and chemical substance, it is used in many industries **(Abdel-Khalek *et al.*, 2015)**, Such as electronics, cosmetics, catalysts, ceramics, dyes, glass, and rubber **(Wang *et al.*, 2004; Ma *et al.*, 2013)**

The toxicity of metal oxide nanoparticles has been reported in freshwater, crustaceans, *Daphnia magna,* and zebra fish, where the effects of exotic organisms are shown in living organisms **(McLoughlin *et al.*, 2000)**. And, **Handy *et al.*, (2008)** showed that toxic nanoparticles of zinc oxide are released into the environment through liquid waste, and the toxic effect appears on vertebrates, including fish.

All changes that occur in Nile tilapia by ZnO-NPs be improved by parsley leaves and parsley nanoparticles as in this study. Parsley has many medical benefits, including anti-bacterial, antioxidant, and protection against liver disease **(Farzaei *et al.*, 2013). Farag *et al.*, (2022)** showed the role of parsley on Nile tilapia against toxicity and it is the role in modulation, prevention, and treatment of the effect of environmental toxins in the aquatic organisms.

Half lethal concentration (LC_{50}):

To examine the toxicity effect, the LC_{50} rating was used as one of the important factors for this study. And, **Prentera *et al.* (2004)** found that Certain levels of pollutants are determined by LC_{50} values. In the current study, the LC_{50} of ZnO-NPs for Nile tilapia, *Oreochromis niloticus,* was determined to be (80 mg/L) and the diameter was (25–35 nm). Meanwhile, **Khodabakhshi *et al.*, (2017)** recorded that the LC_{50} value for the chemical composition of zinc oxide particles was 0.75 mg/L. In rainbow trout. The above observation was supported by **Chakeri *et al.*, (2015)** who noted that heavy metals cause death for these types of fish have different sensitivity to heavy metals. Also, age and size of the fishs can affect on determined LC_{50} **(Rathore and Khangarot., 2002).**

To assess the tolerance and safety levels for a specific contaminant in fish, LC_{50} values are determined **(Prentera *et al.*, 2004)**. In this experiment, LC_{50} values indicated that ZnO-NPs are toxic to tilapia. This result is in agreement with **Alkaladi *et al.*, (2020) and Aziz *et al.,* (2020)**, Differences in fish species may cause differences in LC_{50} values. The difference between the various strains is one of the factors leading to the difference in values of LC_{50} **(Barbieri and Fereira 2011)**. Also, If in the same type of fish, the difference in the LC_{50} value may depend on body weight, age, sex, feeding conditions for the experiment, and water temperature **(Hedayati *et al.*, 2012)**.

Clinical and behavioral changes:

Ghazala *et al.*, (2014) recorded that the behavioural responses are important tools in toxicological studies. In this study, tilapia fish were exposed to LC_{50} ZnO-NPs 80 ml/g, and it affected the external phenomena of fish such as loss of appetite, skin change to dark color, difficulty breathing, and unresponsiveness in escaping quickly, **Alkaladi *et al.*, (2020)**. **Scott and Sloman (2004)** showed that toxic substances affect the metabolism as well as the behavior of fish.

As a result of exposure to zinc oxide, the movement becomes abnormal. These are results supported by **Kaya *et al.*, (2016)** who recorded a changes in the in behavior of tilapia (*O. niloticus*) exposed to ZnO-NPs.

Some studies indicate the toxicity of zinc oxide when used bulk counterparts produces smaller particles, increases surface reaction, and physical and chemical properties changing **(Elsaesser and Howard, 2011)**. Therefore, there is efficiency for the penetration of molecules or their entry into living organisms (**Xiong *et al.*, 2011).**

From this study, we noticed that the co-supplementation of Nile tilapia with parsley leaves and parsley nanoparticles declined the abnormal behavior caused by the exposure fish to ZnO-NPs. This may be due to its protective effects **(Khalil *et al.*, 2015).** Which was more evident in fish fed on parsley nanopartcles (group VI).

Biochemical parameters:

Serum ALT and AST level:

Syuichi and Yone (1978) showed that an increase in levels of AST and ALT is associated with the destruction of the Liver. **Abdel-Tawwab and Wafeek (2017)** assessed health Situations, normal and adaptation to the environment biochemical studies. **(Samipillai and Jagadeesan, 2005)**. To diagnose the liver, AST, ALT, and ALP are evaluated for animals exposed to toxic nanoparticles. **Esmaeillou *et al.*, (2013)** recorded that When mice were exposed to elemental xenon (20-30 nm) at a concentration of 40 mg/ml, caused liver damage. This support the present result which stated that the Levels of ALP, ALT, and AST of Nile tilapia, (*O. niloticus)* exposed to ZnO-NPs increased significantly after 6 weeks of exposure to 8 mg/L ZnO-NPs, **Abdel-Dayem, and Abdeen, (2018)** showed that the release of ALT and AST in the blood when the membrane increased the membrane fluidity and permeability.

Taheri *et al.*, (2017) recorded that the levels of AST, ALT, and ALP increased significantly in cyprinus carpio exposed to zinc oxide for 21 days. Also, **Farsania *et al.*, (2017)** showed that tilapia exposed to 1.5 mg/L and 2.5 mg /L Zinc oxide; increased especially at the higher concentration ALT, ALP, and AST.

These increases are supported by **Alkaladi (2015)** found that ALT, AST, and ALP increased who *O. niloticus* exposed to zinc oxide (1 mg/l, 2 mg/l) for 7 days and the increament was greater after 15 days. **Howe *et al.*, (2013)** explained that this in elevation is due to damage of the liver, which leds to an increase in these enzymes in the blood circulation

Serum urea and creatinine:

(Maita *et al.*, 1984) recorded that to identify glomerular filtration and renal dysfunction, serum (urea, creatinine, and uric acid) is used as an indicator. **(Maxine and Benjamine, 1985)** exhibited that abnormal kidney excretion indicates disorders in fish

As shown in this study the results of urea, creatinine, and uric acid were significantly increased after exposure to 8 mg/L ZnO-NPs for 6 weeks. These results agree with **Abdel Azim *et al.*, (2018)** who found that when fish exposed to at 50 mg /L, kidney function creatinine and urea were increased.

These results are supported by **Srivastav *et al.*, (2016)** who recorded that at doses of ZnO-NPs (300 and 2000) mg/kg in healthy female wistar rats and excretory samples were taken after 24, 48 hours, and 14 days, creatinine was elevated at high dose.

In the same respect, **Gantedi and Anreddy (2012)** when group of mice was exposed to zinc oxide nanoparticles at a dose of 1 or 5 mg/kg body weight for 1 day, 1 week, and 1 month, creatinine level increased after 1 week, but at 1 day and 1 month after exposure, there were no significant changes in serum creatinine levels.

Serum total proteins, albumin, and globulin:

Proteins have an active role in metabolism and water balance **(Bakhshwan *et al.*, 2008)**. Also, protein have an important role in the osmotic balance between the tissue membranes and the circulating blood **(Kadry *et al.*, 2012)**.

Results of total protein, albumin, and globulin to showed a significant ($P \leq 0.05$) increase in tilapia fish exposed 8 mg/L ZnO-NPs However, the A/G ratio markedly significant decreased. In the same respect, **Andrawes *et al.*, (2021)** recorded an increase in total proteins, albumin, and globulin when catfish were exposed to Ag-NPs (50 mg/L) for 30 days. Also, the increase in total proteins were supported by **Manjula *et al.*, (1986)** After exposure to red swamp crayfish, *Procambarus clarkii*, for 14 days to cadmium for 14 days.

The increase in total protein is associated with liver damage, which leads to liver weakness and damage to its cells. This is in agreement with **Öner *et al.*, (2008)** recoded a significant increase in total proteins of Nile tilapia exposed to silver (4.35 mg/L) for 5 days.

Serum glucose level:

An indicator of a stress response is glucose blood **(Abdel-Baky, 2001)**. As shown in the present study serum glucose recorded a significant increase in exposed fish to 8 mg/L ZnO-NPs. These results are in agreement with **Lee *et al.* (2014)** who recorded that glucose level was significantly increased when common carp were exposed to ZnO-NPs (2.4 mg/L) for 12 weeks. This is similar to **Firat and Kargin (2010)** who found an increase in blood glucose level of *O.niloticus* exposed to Zn, Cd, and Zn^{+}, Cd^{+} for 14 days. Environmental pollution and stress on fish may result in high glucose in the blood, which is the result of inhibiting the process of gluconeogenesis and high glycogen breakdown, so glucose in the blood elevated **(Shaluei *et al.*, 2013)**. And, **Hedayati *et al.*, (2018)** who explained that hyperglycemia is due to liver or kidney damage, glycogenolysis, as well as, inhibition of cholinesterase enzyme at neuro-effector site in the adrenal medulla, which leads to the breakdown of glycogen into glucose as a result of excess of secretion cortisol and adrenaline.

In addition, **Reddy and Leatherland (1988)** showed as a result of the increase in glucose in the blood, the cells of the pancreas are injured, which leads to a decrease in insulin and the breakdown of glycogen in the muscles. High blood glucose can also be caused by the breakdown of glycogen in fish that are stressed to need energy **(Abdel-Tawwab and Hamed, 2018)**. But, in different with **Kovacik *et al.* (2018)** who recorded the glucose contents were significantly lowered in the spring season when the freshwater fish C.carpio was exposed to trace metal concentrations.

Serum cortisol level:

Barton (2002); Shaluei *et al*., (2012) stated that the stress-intensity indicator in fish is high level of cortisol in the blood according to

After exposure to toxic metals, cortisol increased in plasma, and this is what most studies have shown **(Laflamme *et al*., 2000)**. The present results showed a significant increase in cortisol in exposed Nile tilapia exposed to 8 mg/L ZnO-NPs after 6 weeks. The present result was agreed with **Johari *et al*., (2013)** who recorded when rainbow trout exposed to colloidal Ag-NPs at nominal concentrations of 100, 32, 10, 3.2, 1, 0.32, 0.1, and 0.032 mg/L, the level of cortisol was increased. Increased cortisol as a result of stress, caused activation of hypothalamic-pituitary interrenal axis, so which led to increase cortisol in blood.

In the same manner, **Bernier and Peter (2001)** who demonstrated that the response to most forms of stress, the HPI axis is activated in the hypothalamic-pituitary-interrenal (HPI), and the end product of this activity is cortisol, which is involved in the regulation of food intake. This disagreed with **Roshanai *et al*., (2012)** who recorded decline in cortisol level of rats exposed to a solution of nano-silver (85 ppm, 65 ppm, 35 ppm, 20 ppm, 5 ppm). This different because difference may be due to different animal species or exposure period.

Serum cholesterol and triglycerides level:

Additionally, the present study recorded no significant in serum levels of Triglycerides, but increase in level of cholesterol demonstrated in Nile tilapia, *O. niloticus* exposed to 8 mg/L ZnO-NPs. These results are in agreement with **Fahmy *et al.*, (2014)** who recorded increase in cholesterol levels of freshwater snail (*Biomphalaria alexandrina*) exposed to ZnO-NPs. While, **Lee *et al.*, (2014)** showed that no change in the level of cholesterol and triglycerides in young common carp, *C. carpio*. In supporting to the previous results, **Öner *et al.*, (2008)** stated that disorder of fat metabolism can be one of the reasons for the increase of cholesterol in the body.

Vajargah *et al.*, (2018) demonstrated that cholesterol analysis is used as an indicator of liver dysfunction, and it is the most sterol that causes a rise in lipid in blood. Other result by **Kovacik *et al.*, (2018)** showed that when *Cyprinus carpio* was exposed to heavy metals, a significant increase in cholesterol in the blood was noted fishes require extra energy, in response to poisoning, leads to an increase in cholesterol levels **(Hamed and Osman, 2017)**.

Similarly, **Vajargah *et al.*, (2018)** who recorded when goldfish (*Crassius auratus gibelio*) exposed to silver nanoparticles, no change in triglycerides. But, **Yu *et al.*, (2014)** recorded that when mice were exposed to TiO_2 nanoparticles by nasal instillation, increased levels of triglycerides in the blood. These results are disagreed with **Canli *et al.* (2017)** who noted that the triglyceride level was decreased in female rats (*Rattus norvegicus*) were exposed to Al_2O_3, CuO and TiO_2

nanoparticles (NPs) for 14 days. This variane may be attributed to the difference in animal species or exposure period.

Oxidative stress and antioxidant enzyme:

One of the indicators that reflect exposure to pollutants and their toxicity on living organisms is antioxidant enzymes, which indicate the antioxidant status of living organisms **(Valavanidis *et al.*, 2006)**. Oxidative stress causes oxidative damage that is related to environmental toxicity **(Velisek *et al.*, 2011)**.

Liver GSH activity:

Organisms have a protective system against the harmful effects of activated reactive oxygen species (ROS) **Jin *et al.*, (2010)**. GSH is a sulfhydryl compound, which has a biological and protective role from toxicants through the SH-group and hydrogen donor of GPx catalyzing the reduction reaction of organic peroxides or catalyzing H_2O_2 **(Moreno *et al.*, 2005)**.

The present results demonstrated a marked lines decrease in GSH content in Nile tilapia exposed to 8 mg/L ZnO-NPs for 6 weeks. These results are in agreement with **Suman *et al.*, (2015)** who recorded a significant decrease in Marine algae, *Chlorella vulgaris, (C. vulgaris)* exposed to ZnO-NPs (>100 mg/L. It is possible that the low GSH is due to the consumption of a large amount of GSH in the detoxification process, and this decrease can lead to the destruction of the antioxidant structures and the accumulation of ROS.

This is supported by **Hao and Chen, (2012)** recorded that after 14 days, (*C.carpio*) which exposed to ZnO-NPs (0.5, 5.0 and 50.0 mg/L), a decrease in GSH was observed. Also, **Li *et al.*, (2008)** observed that GSH was inhibited by medaka exposed to (50.0 mg/L) nano-iron for 14 days.

Whereas, **Xiong *et al.*, (2011)** recorded that the increase in GSH content of zebrafish exposed to ZnO-NPs (5.0 mg/L) may attributed to varied tested chemicals and different exposure periods.

Liver CAT content enzyme:

CAT content is one of vital indicator of oxidative stress is CAT content **(Abdel- Tawwab and wafeek 2017)**. It is also considered among the strong signs of early detection of oxidative environmental pollution **(Abdelazim *et al.*, 2018)**.

The results in this study demonstrated a decrease in liver CAT content in Nile tilapia exposed to ZnO-NPs. **Moreno *et al.*, (2005)** stated that the decrease in CAT is an indicator for oxidant eradication. In agreement with these results **Srikanth *et al.*, (2015)** noted a decrease in CAT, when Chinook salmon cells were exposed to TiO_2-NPs (10-60 μg/ml).

In the same trend **Afifi *et al.*, (2016)** observed a decline in CAT of the brain of *O. niloticus and Tilapia zilli* exposed to Ag-NPs (10-60 mg/L). In support of the result **Jayaseelan *et al.*, (2014)** reported a decrease in the CAT content of Mozambique tilapia, *Oreochromis mossambicus* after exposure to Ni-NPs for 14 days. Nanoparticles have oxidative power by decreasing cells antioxidant system and also producing (ROS) (**Marisa *et al.*, 2016**).

Liver SOD content:

It is the abbreviation for superoxide dismutase, which increases in tissues that suffer from oxidative stress **(Sakai *et al.*, 1992)**. It has a role in balancing oxidants and antioxidants and defending against toxicity **(Li *et al.*, 2003)**. It analyzes free radical superoxide and converts to hydrogen peroxide (H_2O_2) **(Arimoto *et al.*, 2005)**.

From the current study, Nile tilapia was exposed to ZnO-NPs 8 mg/L, the results showed a decrease in SOD content; similar result has also been noted by other authors **Aziz *et al.* (2020)** who recorded decrease in SOD activity in *Labeo rohita* after exposure to ZnO-NPs after 40, 60 and 80 days. Decreased SOD activity leads to cell damage due to the accumulation of reactive oxygen species (ROS) **(Phull *et al.*, 2018)**. This is agreement with **(Zhao *et al.*, 2013)** who found decrease in SOD activity when exposure of embryo-larval zebrafish to ZnO-NPs of 100 mg/L led to an accumulation of ROS in the cell. In supporting to these results **Xiong *et al.* (2011)** recorded that when adult zebrafish were exposed to ZnO-NPs, a significant decrease in the level of SOD in the liver was observed.

Disagreement of these results showed by **Hamed and Osman (2017)** who recorded increased in African catfish *(C. gariepinus)* after exposure to carbofuran. Also, **Benavides *et al.* (2016)** who found that *Carassius auratusafter*, increase in SOD level, after exposure to aluminum (Al2O3) and (ZnO) oxide nanoparticles.

Liver TAC content:

It was showed that Nile tilapia (*O. niloticus*) exposed to 8 mg/L ZnO-NPs had increased total antioxidant capacity (TAC). Oxidation and production of reactive oxygen species due to the presence of metals and nanoparticles **(Rajkumar *et al.*, 2015)**. The present study was supported by **Thwala *et al.* (2013)** found increase in TAC in duckweed (*Spirodela punctuta*) after exposed to ZnO- NPs for 14 days.

In the same line **Andrawes *et al* (2021)** found that an increase in TAC was observed in the liver and kidney tissues of catfish after exposure to 50 mg/L Ag-NPs after 30 days. Increased exposure to toxic chemicals, increases the activity of biomolecules in defense of free radicals **(Choi *et al* 2010)**.

This disagrees with **Khayal *et al.* (2021)** who studied that exposed to ZnO-NPs caused decrease in TAC of male albino rats. Also, **Ibrahim *et al.* (2021)** found that decrease in TAC of mice after exposed to ZnO-NPs. This agrees with **Abbasalipourkabir *et al.* (2015)** recorded increase in TAC in adult male Wistar rats after exposed to ZnO-NPs.

Liver GPx content:

It plays an important role in getting rid of toxic substances that will be destroyed in mitochondria in the presence of GSH **(Doyotte *et al.*, 1997)**. It also works to detoxify hydroperoxides **(Orbea *et al.*, 2000)**.

In the current study, Nile tilapia was exposed to ZnO-NPs 8 mg/L, the results showed a decrease in GPx content. this study is in agreement with **Ali *et al.* (2012)** who recorded a decrease in GPx when *Lymnaea luteola* (*L. luteola*) was exposed to ZnO-NPs. ZnO-NPs led to an increase in the production of ROS, especially O2, so large amounts of GSH were depleted and led to a decrease in GPx.

This is in agreement with **Chandran *et al.* (2005)** who recorded that when *A. fulica* snails were exposed to zinc, led to a decrease in GPx level in the digestive gland and kidney.

Liver NO content:

Nitric oxide (NO) is one of the key mediators of immune defense, a reactive form of nitrogen, and plays an important role in cellular signaling molecules. NO is one of the free radicals that work to maintain the balance of innate immunity in combination with the production of inflammatory mediators needed by macrophages leading to the elimination of extraneous particles and pathogens **(Kumar *et al.* 2018)**

In this study, *O. niloticus* exposed to 8 mg/L ZnO-NPs had decreased NO. **Xia *et al* (2008)** reported that ZnO nanoparticles induce oxidative stress and apoptosis in macrophages and its effects on viability, phagocytic activity, and NO production. It was agreement **Kumar *et al.* (2018)** reported that when Balb/c mice were exposed to ZnO-NPs.

This disagrees with **Fahmy *et al.* (2014)** reported that the freshwater snail (*Biomphalaria alexandrina)* was exposed to ZnO-NPs. And, for 6 consecutive days, freshwater bivalve mollusk *Coelatura aegyptiaca* was exposed to ZnO-NPs **(Fahmy and Sayed 2017)**. The same year, **Wang *et al.* (2017)** found that NO was increased, when exposed to rats at different doses ZnO-NPs (200, 400, and 800 μg /kg).

Liver MDA content:

MDA measured indicates the effect of ZnO-NPs that caused oxidative injury **(Nusetti *et al.* 2001)**. Increased levels of MDA are indicative of increased ROS due to toxicity of ZnO-NPs **(Syama *et al.* 2013)**

In this study, that Nile tilapia (*O. niloticus*) exposed to 8 mg/L ZnO-NPs had increased MDA. It is agreement with **Dieni *et al.* (2014)** found that an increase in MDA was observed, when *C. commersonii* was exposed to ZnO-NPs. And, when *Oncorhynchus mykiss* was exposed to ZnO-NPs **(Connolly *et al.* 2016)**. Also, Increased MDA was observed in zebrafish when exposed to ZnO-NPs **(Xiong *et al.* 2011)**.

MDA is a strong tissue predictor of oxidative stress. ZnO-NPs are one of the substances that lead to the generation of oxidative stress **(Abdelazim *et al.* 2018)**, it is agreement with **(Mir *et al.* 2020)** reported that exposure to ZnO-NPs increases MDA, as a result of oxidative stress.

Blood parameters:

Changes in the blood that indicate a response to environmental changes **(Hawkins and Mawdesley, 2006)**.

In this study, that Nile tilapia (*O. niloticus*) exposed to 8 mg/L ZnO-NPs had no change in RBCs and WBCs, but increase in HB. The increase in haemoglobin is due to a simultaneous decrease in oxygen and a rise in carbon dioxide. It is agreement with **(Montgomery *et al.* 2018)**.

Also, in this study, insignificant in MCV, but, increase MCHC, it is agreement with **(Aliakbar and Zahra 2013)** and decrease MCH were observed it agreement with **(Ghalazy 1992)** reported that MCH was decrease as a result of poor oxygen absorption as a result of destruction in the gills due to exposure to ZnO-NPs

White blood cells have a role in defending fish. They contain granulocytes, monocytes, lymphocytes, and platelet cells. Granulocytes and monocytes act as phagocytes to rescue debris from tissues **(Ellis et al. 1978; Maheswaran *et al.* 2008)**

Lympho, Basophils, Neutro and Eosinophil insignificant were observed. But Monocyte was decrease it is agreement with **(Van der Merwe, 1992)** found that reducing of monocytes in O. *mossambicus* in affected tissues such as gills, liver and kidneys that were damaged by copper.

The platelets were decrease, it may indicate an obstruction of platelets in the spleen or a lack of platelet production **(Devineni *et al.* 2015).**

Evaluation of serum immunoglobulin:

Measurement of immunoglobulin G (IgG) and M (IgM) levels in the blood indicating the effect of metal oxide nanoparticles on the immune system. Reducing the level of immunoglobulin indicates the an immunotoxic effect of metal oxide nanoparticles **(Kargin 2021)**, In this study, that Nile tilapia (*O. niloticus*) exposed to 8 mg/L ZnO-NPs had decreased IgG, Similar to the presented study results, **(Vandebriel *et al.*2014)**

But, in this study, that Nile tilapia (*O. niloticus*) exposed to 8 mg/L ZnO-NPs had increased IgM. It is the first antibody to appear in response to initial exposure to an antigen. It is agreement with **Jacobsen *et al.* (2015)**

Histology:

The microscopic examination of the liver of tilapia intoxicated with zinc oxide revealed severe vascular changes in the form of congestion, thrombosis, focal areas of hemorrhages around the dissociated and necrotic pancreatic acini and hemolysis of erythrocytes within the portal veins with mononuclear inflammatory cellular infiltration. these changes could be attributed to the effect of zinc oxide on the endothelium of the blood vessels. In addition to these vascular changes, hepatic damage represented by degeneration, necrosis and apoptosis of hepatic cells were also recorded. these histopathological findings were reflected on the results of serum biochemical analysis where both AST and ALT values were significantly increased compared with control group. Moreover, bile ductal hyperplasia and periductal inflammatory cellular infiltration were found in tilapia intoxicated with zinc oxide. This change could be attributed to the irritant effect of zin oxide on the lining epithelium of the bile duct and confirmed by serum analysis where the ALP levels were elevated compared with control group.

The histopathological examination of liver of tilapia intoxicated with zinc oxide and treated with parsley revealed improvement in the microscopic picture of the liver where congestion with mononuclear inflammatory cells aggregation and necrosis of few hepatocytes and dissociation of the pancreatic acini were recorded. These improvement in the microscopic picture of the liver was confirmed by the serum analysis where the levels of AST, ALT and ALP were restored and became nearly similar to those of control group

Also, the histopathological examination of liver of tilapia intoxicated with zinc oxide and treated with parsley nanoparticles revealed enhancement in the microscopic picture of the liver comparing with the examined liver of tilapia intoxicated with zinc oxide alone. Comparing with group of tilapia intoxicated with zinc oxide and treated with parsley, the examined liver of this group showed greater improvement where congestion of and focal aggregation of few mononuclear inflammatory cells with necrosis of hepatocytes and dissociation of some pancreatic acini were seen in few examined liver. These findings were coming in parallel with the results of serum analysis where AST, ALT and ALP levels were comparing with those of tilapia intoxicated with zinc oxide and treated with parsley

In the present investigation, the microscopic examination of the liver of tilapia intoxicated with ZnO-NPs revealed severe vascular changes in the form of congestion, thrombosis, focal hemorrhages around the dissociated and necrotic pancreatic acini and hemolysis of erythrocytes within the portal vessels with mononuclear inflammatory cellular infiltration of the hepatic parenchyma. These changes could be attributed to the effect of ZnO-NPs on the endothelium of the blood vessels. In addition to these vascular changes, hepatic damage represented by degeneration, necrosis and apoptosis of hepatic cells were also recorded. These findings were confirmed by **Hao *et al.*, (2013) Shahzad *et al.*, (2018) Harsij *et al.*, (2021) and Khan**

***et al., (*2022)** who reported nearly similar microscopic changes in the liver of different fish species exposed to ZnO-NPs. In our study, the recorded histopathological changes in the liver were reflected on the result of serum biochemical analysis where both AST and ALT values were significantly increased compared with control group. These findings were supported by **Syuichi and Yone (1978)** who mentioned that destruction of hepatic cells was associated with increase in AST and ALT levels. In addition, elevations in serum AST and ALT levels were also recorded after exposure to zinc oxide in Cyprinus carpio (**Taheri et al 2017),** and Tilapia (**Alkaladi, 2015; Farsania *et al.*, 2017**).

In the present investigation, bile ductal hyperplasia and periductal inflammatory cellular infiltration were found in the examined liver of tilapia intoxicated with ZnO-NPs. These changes could be attributed to the irritant effect of ZnO-NPs on the lining epithelium of the bile duct and confirmed by increases in the serum ALP levels compared with control group. This opinion was supported by **Moss (1997**) who mentioned that, hepatic ALP is present on the surface of bile duct epithelia and cholestasis enhances the synthesis and release of ALP, and accumulating bile salts increase its release from the cell surface. Therefore, serum ALP levels usually rise in bile duct obstruction. In addition, **Alkaladi (2015)** and **Taheri *et al.*, (2017)** found that serum ALP values were increased in *O. niloticus* and *C.carpio* after exposure to zinc oxide.

Regarding to evaluation of the effects of parsley leaves and parsley nanoparticles on the liver of tilapia after exposure to ZnO-NPs. The histopathological examination of liver of tilapia intoxicated with ZnO-NPs and treated with parsley leaves revealed improvement in the microscopic picture of the liver where congestion with mononuclear inflammatory cells aggregation and necrosis of few hepatocytes and dissociation of the pancreatic acini were recorded. These improvement in the microscopic picture of the liver was confirmed by the serum analysis where the levels of AST and ALT were restored but still higher than those of control group. In addition, serum ALP values were nearly similar to those of control which confirmed the absence of the pathological changes in the bile ducts in liver of tilapia after exposure to ZnO-NPs and treated with parsley leaves.

Also, the histopathological examination of liver of tilapia intoxicated with zinc oxide and treated with parsley nanoparticles revealed enhancement in the microscopic picture of the liver comparing with the examined liver of tilapia intoxicated with zinc oxide alone. Comparing with group of tilapia intoxicated with zinc oxide and treated with parsley, the examined liver of this group showed greater improvement where congestion of and focal aggregation of few mononuclear inflammatory cells with necrosis of few hepatocytes and dissociation of some pancreatic acini were seen in few examined liver. These findings were coming in parallel with the results of serum analysis where AST and ALT levels were lower than those recorded in tilapia intoxicated with zinc oxide and treated with parsley leaves and the recorded values were nearly similar to those of the control group

Parsley:

Farzaei *et al.*, (2013) recorded that parsley is one of the Umbelliferae family, originated in the Mediterranean, and is currently cultivated in several places around the world.

From the present study, it was determined that increase in liver enzymes and kidney function in Nile tilapia expose to ZnO-NPs, indicating hepato-renal injury. These results may be attributed to ZnO-NPs accumulation and deposition in fish tissues. As expected, feeding fish with dietary parsley leaves and parsley nanoparticles diet reduced liver enzymes (ALT, ALP, and AST) and kidney function (urea, creatinine) levels, demonstrating its protective effect against ZnO-NPs. In a similar study **Khalil *et al.*, (2015) ; Rahmat *et al.*, (2018)** noted the effect of parsley on liver and kidney of rat.

Parsley has numerous beneficial properties, including its ability to improve antioxidant parameters (LPO, SOD, CAT, NO, GPx and MDA), it is in agreement with **Wong and kitts (2006)**.

Parsley supplementation has been shown to reduce disease such as heart diseases, urinary tract, skin diseases, high blood pressure, hepatotoxicity and diabetes **Farzaei *et al.*, (2013)**. parsley also contains antioxidants, anti-inflammatory and immune stimulants (**Hassan *et al.*, 2004; Abdul Latif *et al.*, 2017**).

Conclusion:

In conclusion, based on the results obtained from the present study, it was found that exposure to 8 mg/L ZnO-NPs and/ or NS produced significant modulator effects on the varies behavioural and clinical changes of catfish which presenting as morphological alterations after exposed to ZnO-NPs.

The decline in the liver function (ALP, ALT and AST) and kidney function (Urea and creatinine) under the effect of ZnO-NPs and /or parsley might have been the result of a direct action on a damage occurred in the hepatic and renal tissues. In addition, the decrease in total protein, glucose, cortisol level and by ZnO-NPs and NS administration may be related to the raise in glucogenolysis and liver damage.

The current study showed that administration of parsley stimulated the total lipid, oxidant and antioxidant of Nile tilapia and improved them to levels similar to the control group.

Further studies are still required to understand the toxicological mechanisms of ZnO-NPs in fishes so that coming studies should shed new lights on the important activity of ZnO-NPs ultimately biochemical, histo- pathological and other adverse impacts.

Summery

SUMMARY

Physiological, histological, and immunological study on the effect dietary parsley leaves extract and parsley nanoparticles against Zinc oxide nanoparticles toxicity in Nile tilapia, *Oreochromis niloticus*

The present work has been planned to investigate the effect of ZnO-NPs on selected biochemical parameters of serum, and liver tissue in *O. niloticus*. This study was carried out also to evaluate the effect of ZnO-NPs on haemato-biochemical parameters oxidative stress, and antioxidant enzymes, In addition, this study aimed to highlight the histopathological change to ZnO-NPs exposure. Finally, to represent the role of parsley leaves and parsley nanoparticles in the improvement of these parameters.

Experimental design and investigated parameters:

In the present study, 400 Nile tilapia (*O. niloticus*) were used. fifty fish were used to estimate the (LC_{50}) of ZnO-NPs. The other 180 fish were divided into six groups and each group consisted of 3 replicates of 10 fish per aquarium. Group I: served as the control group. Group II: served as the control group and co-administrated with parsley at the rate of 4g/kg diet. Group III: served as the control group and co-administrated with nano parsley at the rate of 4mg/kg diet Group IV: exposed to 8 mg/L ZnO-NPs in deionized water Group V: exposed to 8 mg/L ZnO-NPs and fed on 4g/kg parsley. Group VI: exposed to 8 mg/L ZnO-NPs and fed on 4g/kg diet nano parsley.

After 6 weeks day of the experiment, blood samples were collected from the caudal blood vessels. Serum was collected, liver and kidney were rapidly excised for the biochemical and histological studies.

The present work included the following investigations:

- Clinical investigation and post-mortem examination
- Serum: Determination of serum levels of ALP, ALT, AST, urea, creatinine, total proteins, Albumin, Globulin, glucose, cortisol, cholesterol, and triglycerides.
- Liver and kidney determination of LPO, GSH, CAT, SOD

TAC, NO, MDA and GPx content.

- Liver histopathological changes.

The results:

The half-lethal concentration (LC_{50}) of ZnO-NPs was observed through exposure to Nile tilapia (*O. niloticus*) was 8 mg/L after 96 h.

Clinical behavior and biochemical parameters:

The clinical signs of *O. niloticus* associated with toxicity of 8 mg/L ZnO-NPs were manifested by abnormal skin pigmentation, moving towards the air pump, staying for a long time, a certain location, also irregular movement, loss of equilibrium, anorexia, detached scales, frayed fins, and Postmortem lesions showed lipid degeneration in the liver with paleness in other internal organs whereas observed.

The postmortem findings of exposed fish were different degrees crowding especially in all internal organs of the exposed fish. As well as, the enlargement of liver. The biochemical analysis revealed an increase in liver enzymes values (ALP, ALT and AST). Also, urea, creatinine, uric acid, total protein, albumin, globulin glucose, cortisol and total lipid levels were also detected in group (IV) exposed to 8 mg/L ZnO-NPs for 6 weeks days.

Oxidative stress parameters in liver tissues recorded that:

The exposure of 8 mg/L ZnO-NPs in Nile tilapia (*O. niloticus*) led to increase hepatic total antioxidant capacity (TAC) and Malondialdehyde (MDA), where reduced glutation (GSH), catalase (CAT), superoxide dismutase (SOD), Nitrous oxide (NO) and Glutathione Peroxidase (GPx)

The most observed histopathological alterations included:

-Liver: Hepatocytes with hydropic degeneration, portal tract showing inflammatory cellular infiltrate, mild fibrosis, interface hepatitis, and shows multiple foci of spotty necrosis.

Conclusion:

Feeding of *O. niloticus* on parsley leaves and/or parsley nanoparticles enriched diet minimized the side effects of ZnO-NPs toxicity and improve the above-mentioned variables near to the control group. The current study has revealed that dietary parsley leaves extract and parsley nanoparticles have a good effect against ZnO-NPs toxicity in Nile tilapia (*O. niloticus*).

References

REFERENCES

Abass, M. A.; Selim, S. A.; Selim, A. O.; El-Shal, A. S. and Gouda, Z. A. (2017). Effect of orally administered zinc oxide nanoparticles on albino rat thymus and spleen. IUBMB life.,68: 528-538.

Abbasalipourkabir, R.; Moradi. H.; Zarei, S.; Asadi. S.; Salehzadeh, A.; Ghafourikhosroshahi, A.; Mortazavi, M. and Ziamajidi, N. (2015). Toxicity of zinc oxide nanoparticles on adult male Wistar rats. Food Chem. Toxicol.,84: 154-160.

Abd El-Atti, M.; Desouky, M.; Mohamadien, A. and Said, R. (2020). Sub-lethal toxicity of silver nanoparticles on crayfish, *procambarus clarkii*: oxidative stress, DNA damage, and histopathological alterations. Arct. J., 73(3): 78-108.

Abdelazim, A.M.; Saadeldin, I. M.; Swelum, A.A.; Mohamed, M.A.; Afifi, M. M. and Alkaladi, A. (2018). Oxidative Stress in the Muscles of the Fish Nile Tilapia caused by Zinc Oxide Nanoparticles and Its Modulation by Vitamins C and E. Oxid. Med. Cell. Longev.

Abdel-Baky, T. E. (2001). Heavy metals concentrations in the catfish, *Clarias gariepinus* (Burchell, 1822) from River Nile, El-Salam canal and Lake Manzala and their impacts on cortisol and thyroid hormones. Egypt. J. Aquat. Biol. Fish., 5(1) :78-88.

Abdel-Daim, M.M.; Abdeen, A.; (2018). Protective effects of rosuvastatin and vitamin E against fipronil-mediated oxidative damage and apoptosis in rat liver and kidney. Food Chem.Toxicol.114:68–77.

Abdel-Daim, M.M; Eissa, I.A.M.; Abdeen, A.; Abdel-Latif, H.M.R.; Dawood, M.A.O.; and Hassanf, A.M. (2018). Lycopene and resveratrol ameliorate zinc oxide nanoparticles-induced oxidative stress in Nile tilapia, *Oreochromis niloticus*. Environ. Toxicol. Pharmacol.68:44-50.

Abdel-Khalek, A.A.; Kadry.M.; Hamed.A. and Marie.M.(2015). ecotoxicological impacts of zinc metal in comparison to its nanoparticles in Nile tilapia; *Oreochromis niloticus*. JOBAZ.,72: 113-125

Abdellatief, S.A.; Galal, A.A.A; Farouk , S.M.; Mohamed M.; and Abdel-Daim, M.M. (2017). Ameliorative effect of parsley oil on cisplatin-induced hepato-cardiotoxicity: A biochemical, histopathological, and immunohistochemical study. Biomed. Pharmacother. 86: 482-481.

Abdel-Tawwab, M.; and Wafeek, M. (2017). Fluctuations in water temperature affected waterborne cadmium toxicity: hematology, anaerobic glucose pathway, and oxidative stress status of Nile tilapia, *Oreochromis niloticus* (L.). Aquaculture. 477: 106-111.

Adhikari, T.; Kundu S.; Biswas A.K.; Tarafdar, J.C. and Rao, A.S. (2012). Effect of copper oxide nano Particle on seed germination of selected crops. J. Agri. Sci. Tech. 2: 815- 823.

Aebi, H. (1984). Catalase in vitro. Methods Enzymol., 105:121-126.

Afifi, M.; Saddick, S.; and Abu Zinada, O. A. (2016). Toxicity of silver nanoparticles on the brain of *Oreochromis niloticus* and *Tilapia zillii*. Saudi J. Biol. Sci., 23: 754–760

Ali, D.; Alarifi, S.; Kumar, S.; Ahamed, M.; and Siddiqui, M. A. (2012). Oxidative stress and genotoxic effect of zinc oxide nanoparticles in freshwater snail, *Lymnaea luteola* L. Aquat. Toxicol. 124-125: 83-80

Aliakbar, H.; and Zahra, G. (2013). Effect of mercuric chloride on some hematological, biochemical Parameters in silver carp *(Hypophthalmichthys Molitrix)*. Int. J. Vet: Research & Reports. 11 pages.

Alkaladi, A. (2018). Vitamins E and C ameliorate the oxidative stresses induced by zinc oxide nanoparticles on liver and gills of *Oreochromis niloticus*. Saudi J. Biol. Sci. 28: 357-362

Alkaladi, A.; Afifi, M.; Ali, H.; and Saddick, S. (2020). Hormonal and molecular alterations induced by sub-lethal toxicity of zinc oxide nanoparticles on *Oreochromis niloticus* Saudi J. Biol. Sci. 27: 1286-1301

Alkaladi, A.; Nasr El-Deen, N. A. M; Afifi, M.; Abu Zinadah, O. A. (2015). Hematological and biochemical investigations on the effect of vitamin E and C on *Oreochromis niloticus* exposed to zinc oxide nanoparticles. Saudi J. Biol. Sci. 22, 556-563.

Altunbaş, M.; and Türel, İ. (2009). *Petroselinum crispum* (maydanoz) tohumu uçucu yağ özütünün letal doz düzeyleri ve antienflamatuvar aktivitesinin deney hayvanları üzerinde araştırılması. Yüzüncü Yıl Üniversitesi Veteriner Fakültesi Dergisi, 20(1): 21-25.

American Veterinary Medical Association (AVMA). (2017). Judicious use of antimicrobials for treatment of aquatic animals by veterinarians.

Amin, N.; Zulkifli, S.Z.; Azmai, M.N.A.; and Ismail, A.(2021) Toxicity of Zinc Oxide Nanoparticles on the Embryo of Javanese Medaka (*Oryzias javanicus Bleeker*, 1854): A comparative Study. Animals ., 11

Amlacher, E. (1970). "Text book of fish diseases " pp:135-137. T. E. S. publication, 1870, Jercy, USA.

Andersson-Willman, B.; Gehrmann, U.; Cansu, Z.; Buerki-Thurnherr, T.; Krug, H.F.; Gabrielsson, S.; and Scheynius, A. (2012). Effects of subtoxic concentrations of TiO2 and ZnO nanoparticles on human lymphocytes, dendritic cells and exosome production. Toxicol Appl. Pharmacol, 264:84–103.

Andrawes, N. G.; Abo Nour, A.A.; Shaheen, A. A. M.; Hamed, H. (2021). Effect of *Nigella sativa* enriched diet on biochemical variables and antioxidant damage caused by silver nanoparticles toxicity in the African catfish, *Clarias gariepinus*. Egypt. J. Aquat. Biol. Fish., 25(3): 63 – 76

Ansar, S.; Abudawood, M.; Alaraj, A.; and Hamed, S. (2018). Hesperidin alleviates zinc oxide nanoparticle induced hepatotoxicity and oxidative stress. BMC Pharmacol. Toxicol. 19(1):65.

Ansar, S.; Abudawood, M.; Hamed, S. S.; and Aleem, M. M. (2016). Exposure to zinc oxide nanoparticles induces neurotoxicity and proinflammatory response: Amelioration by hesperidin. Biol. Trace Elem. Res. 175: 360–366

Arimoto, T.; Kadiiska, M. B.; and Sato, K. (2005). Synergistic production of lung free radicals by diesel exhausts particles and endotoxin. Am. J. Respir. Crit. Care Med. 171:378-87.

Asghar, M. S.; Qureshi, N. A.; Jabeen, F.; Khan, M. S.; Shakeel. M.; and Chaudhry, A. S. (2018). Ameliorative effects of selenium in ZnO NP-induced oxidative stress and hematological alterations in *Catla catla.* Biol. Trace Elem. Res., 186:278–287

Aziz, S.; Abdullah, S.; Abbas, K.; and Zia, M. A.(2020). Effects of engineered zinc oxide nanoparticles on freshwater fish, *Labeo rohita*: Characterization of ZnO Nanoparticles, acute toxicity and oxidative stress. Pak. Vet. J., 40(4): 478-483.

Baladehi, C. E.; Hedayati, S.A.A.; Kolangi Miandare, F. H.; and Bagheri, T. (2017). The hematological improvement of rainbow trout (*Oncorhynchus mykiss*) during dietary supplementation with vitamin C after exposure to zinc nano-particles. IJFS. 16:162- 168

Banaee, M.; Vaziriyan, M.; Derikvandy, A.; Haghi, B.N.; and Mohiseni, M. (2018). Biochemical and physiological effect dietary supplements of ZnO nanoparticles on common carp (*Cyprinus carpio*). Int. J. Aquat. Biol. (7): 56-64

Barbieri, E.; and Ferreira, L. A. A. (2011). Effects of the organophosphate pesticide Folidol 600 on the freshwater fish, Nile tilapia, *Oreochromis niloticus*. Pest. Biochem. Physiol., 88 (3): 208-214.

Barton, B.A. (2002). Stress in fishes: a diversity of responses with particular reference to changes in circulating corticosteroids. Integr. Comp. Biol. 42: 517–525.

Becheri, A.; Durr, M.; Lo Nostro, P.; and Baglioni, P.; (2008). Synthesis and characterization of zinc oxide nanoparticles: application to textiles as UV-absorbers, J. Nanoparticle Res., 10: 678–688.

Beitsayah, A.; Banaee, M.; and Haghi, B. N. (2018). Effects of oral administration of nano zinc oxide on some immunological parameters of common carp (*Cyprinus carpio*). J. Aqu. Eco., 8 (3) :17-27

Benavides, M.; Fernández-Lodeiro, J.; Coelho, P.; Lodeiro, C.; and Diniz, M. S. (2016). Single and combined effects of aluminum (Al_2O_3) and zinc (ZnO) oxide nanoparticles in a freshwater fish, *Carassius auratus*. Environ. Sci. Pollut. Res., 23(24):24578-24591

Benn, T.M.; and Westerhoff, P. (2008). Nanoparticle silver released into water from commercially available sock fabrics, Environ. Sci. Technol. 42: 7025–7026.

Bernier, N. J.; Peter, R. E. (2001). The hypothalamic-pituitary-interrenal axis and the control of food intake in teleost fish. Comp. Biochem. Physiol. B 128: 638-644.

Beutler, E.; Duron, O.; and Kelly, B. M. (1963). Improved method for the determination of blood glutathione. J. Lab. Clin. Med., 61: 882-888.

Bhavya, C.; Yogendra, K.; Mahadevan, K.M.; and Madhusudhana, N. (2016). Synthesis of calcium oxide nanoparticles and its mortality study on fresh water fish *Cyprinus carpio*. IOSR-JESTFT 12: 55–60.

Brand, M.E. (2001). Bioaccumulation of metals in *Labeo Congoro* from the Olifants River (Mpumalanga) and the effect of Nickel on the haematology of fish. (M.Sc. Sci. thesis). Rand Afrikaans Univ.

Campbell, T. W. (2004). Hematology of fish. In: Troy DB, editor. Veterinary hematology and clinical chemistry. Baltimore: Lippincott Williams and Wilkins Premier Journal Collection, 277-288.

Canli, E. G.; Atli, G.; and Canli, M. (2017). Response of the antioxidant enzymes of the erythrocyte and alterations in the serum biomarkers in rats following oral administration of nanoparticles. Environ. Toxicol. Pharmacol. 50:145-150

Carleton, H.M.; Drury, R.A.B.; and Wallington, E.A. (1967). 4th ed. Carleton's Histological technique. Oxford University Press. New York.

Chakeri, R.; Sajadi, M. M.; Kamrani, E.; and Aghajari, N. (2015). Determination of heavy metal (lead and cadmium) concentrations in liver and muscle tissue of Indian mackerel (*Rastrelliiger kanagurta*) in Persian Gulf. Iran. Sci. Fish. J., 24 (2): 115-125.

Chandran, R.; Sivakumar, A. A.; Mohandass, S.; and Aruchami, M. (2005). Effect of cadmium and zinc on antioxidant enzyme activity in the gastropod, *Achatina fulica*. Comp. Biochem. Physiol. Part - C: Toxical. Pharmacol. 140: 422–426.

Chang, Y. N.; Zhang, M.; Xia, L.; Zhang, J.; and Xing, G. (2012). The toxic effects and mechanisms of CuO and ZnO nanoparticles. Materials, 5(12): 2850-2871.

Charles, D. J. (2012). Parsley. Handbook of herbs and spices (2nd ed), (1:430–451).

Chen, T.; Lin, C.; and Meng, P. (2014) Zinc oxide nanoparticles alter hatching and larval locomotor activity in zebrafish (*Danio rerio*) . J. Hazard. Mater.(277): 134-140

Choi, J. E.; Kim, S.; Ahn, J. H.; Youn, P.; Kang, J. S.; Yi, J.; and Ryu, D. Y. (2010). Induction of oxidative stress and apoptosis by silver nanoparticles in the liver of adult zebrafish, Aquat. Toxicol., 100: 151–158.

Chupania, L.; Niksirata, H.; Velíšeka, J.; Staráa, A.; Hradilovább, S.; Kolaříkb, J.; Panáčekb, A.; and Zuskováa, E. (2018). Chronic dietary toxicity of zinc oxide nanoparticles in common carp (*Cyprinus carpio* L.): Tissue accumulation and physiological responses. Ecotoxicol. Environ. Saf. 147: 110-116.

Clark, R. J.; and Menary, R. C. (1983) The tasmanian essential oil industry–production of steamdistilled essential oil crops, 8th International Essential Oil Congress, 14–17 March, Singapore.

Connolly, M.; Fernández, M.; Conde, E.; Torrent F.; Navas J. M.; and Fernández-Cruz, M. L. (2016). Tissue distribution of zinc and subtle oxidative stress effects after dietary administration of ZnO nanoparticles to rainbow trout. Sci. Total Environ., 551: 334-343.

Craig, S.; Kuhn, D. D.; and Schwarz, M. H. (2017). Understanding fish nutration, feeds and feeding. VCES. 420: 256-266.

Crestani, M.; Menezes, C.; Glusczak, L.; Miron, dos S.D.; Spanevello, R.; Silveira, A.; Goncalves, F. F.; Zanella, R.; and Loro, L. V. (2007). Effect of clomazone herbicide on biochemical and histology aspect of silver catfish (*Rhamdia quelen*) and recovery pattern. Chemosphere. 67(11):2305-2311.

Dehkourdi, B.; Fatahian, S.; and Shahanipoor, K. (2017). Synthesis, characterization and renal toxicity of ZnO and polyethylene glycol coated ZnO nanoparticles. Nanomed. J. (1): 55-60.

Dekani, L.; Johari, S. A.; and Joo, H. J.(2018). Comparative toxicity of organic, inorganic and nanoparticulate zinc following dietary exposure to common carp (*Cyprinus carpio*). Sci. Total Environ., 656: 1181-1188.

Deore, M. S.; Keerthana, S.; Naqvi, S.; Kumar, A.; and Flora, S. J. S. (2021). Alpha-Lipoic acid protects co-exposure to lead and zinc oxide nanoparticles induced Neuro, Immuno and male reproductive toxicity in rats. Neuro, immune front. Pharmacol. 12:626238.

Devineni, S. B.; Suneetha, O.; and Harshavardhan, N. (2015) Study of platelet count in malaria patients and the correlation between the presence and severity of platelet count with type of malaria. J. Evol. Med. Dent. Sci. 4(67):11734–11746.

Dieni, C. A.; Callaghan, N. I.; Gormley, P. T.; Butler, K. M. A. and Mac Cormack, T. J. (2014). Physiological hepatic response to zinc oxide nanoparticle exposure in the white sucker, *Catostomus commersonii*. Comp. Biochem. Physiol. Part – C. Toxical. Pharmacol., 162: 51-61.

Dobretsov, S.; Sathe, P.; Bora, T.; Barry, M.; Myint, M.; and Al Abri, M. (2020) Toxicity of different zinc oxide nanomaterials at 3 trophic Levels: Implications for Development of low-toxicity antifouling agents. Environ. Toxicol. Chem.(38): 1343-1354.

Doyotte, A.; Cossu, C.; Jacquin, M.C.; Babut, M.; and Vasseur, P. (1997). Antioxidant enzymes, glutathione and lipid peroxidation as relevant biomarkers of experimental or field exposure in the gills and the digestive gland of the freshwater bivalve, *Unio tumidus*. Aquat. Toxicol. 38: 83–110.

Dumas, B. T.; and Biggs, H. G. (1972). In stander method of clinical chemistry. Academic Press New York.

El-Barbary, M.I.; Mehrim, A.I. (2008). Protective effect of antioxidant medicinal herbs, Rosemary and Parsley, on subacute aflatoxicosis in *Oreochromis niloticus*. J. Fish. Aquac. Sci., 4 (4): 178–180.

Ellis, E. A.; Munroe, A. L. S.; and Roberts, R. J. (1978). Fish Pathology. Bailliere Tindall, London., pp.318.

Elsaesser, A.; and Howard, C.V. (2011). Toxicology of nanoparticles. Adv. Drug Deliver. Rev., 64:128–137

Esmaeillou, M.; Moharamnejad, M.; Hsankhani, R.; Tehrani. A; and Maadi, H. (2013). Toxicity of ZnO nanoparticles in healthy adult mice. Environ. Toxicol. Pharmacol. 35: 67-71.

Esmaeilnejad, B.; Samiei, A.; Mirzaei, Y.; and Farhang-Pajuh, F. (2018). Assessment of oxidative/nitrosative stress biomarkers and DNA damage in *Haemonchus contortus*, following exposure to zinc oxide nanoparticles. Acta Parasitologica.,63:563–571.

Fahmy, S. R.; Abdel-Ghaffar, F.; Bakry, F. A.; and Sayed, D. A. (2014). Ecotoxicological effect of sublethal exposure to zinc oxide nanoparticles on freshwater snail, *Biomphalaria alexandrina.* Arch. Environ. Contam. Toxicol., 67:182–202.

Fahmy, S. R.; and Sayed, D. A. (2017). Toxicological perturbations of zinc oxide nanoparticles in the *Coelatura aegyptiaca* mussel. Toxicol. Ind. Health., 33: 564–575.

Farag, M. R.; Alagawany, M.; Khalil, S. R.; Abd El-Aziz, R.M.; Zaglool, A.W.; Moselhy, A. A. A.; and Abou-Zeid, S. M. (2022). Effect of parsley essential oil on digestive enzymes, intestinal morphometry, blood chemistry and stress-related genes in liver of Nile tilapia fish exposed to bifenthrin. Aquacaluture. 564: 737322.

Farsania, H.G.; Doriab, H.B.; Jamalic, H.; Hasanpourc, S.; Mehdipourd, N.; and Rashidiyane, G. (2017). The protective role of vitamin E on *Oreochromis niloticus* exposed to ZnO-NP. Ecotoxicol. Environ. Saf.,147:1-7.

Farzaei, M.H.; Abbasabadi, Z.; Ardekani, M.R.; Rahimi, R.; and Farzaei, F. (2013). Parsley: a review of ethnopharmacology, phytochemistry and biological activities, J. Tradit. Chin. Med. 33: 815–826.

Fernández, D.; García-Gómez, C.; and Babín, M. (2013). In vitro evaluation of cellular responses induced by ZnO nanoparticles, zinc ions and bulk ZnO in fish cells. Sci. Total Environ. 452-453:262-74.

Firat, O¨.; and Kargin, F. (2010). Individual and combined effects of heavy metals on serum biochemistry of Nile Tilapia, *Oreochromis niloticus*. Arch. Environ. Contam. Toxicol. 58: 151–157.

Foster, L.; and Dunn, R. (1974). Single antibody technique for radioimmunoassay of cortisol in unextracted serum or plasma. Clin. Chem., 20: 365-372.

Fu, W.; Magnúsdóttir, M.; Brynjólfson, S.; Palsson, B. Ø.; and Paglia, G. (2012). UPLCUV-MSE analysis for quantification and identification of major carotenoid and chlorophyll species in algae. Anal. and Bioanal. Chem., 404(10):3145-54.

Gantedi ,S.; and Anreddy, R. (2012). Toxicological studies of zinc oxide nanomaterials in rats. Toxicol. Environ. Chem.1768-1778.

Ghalazy, K. S. (1992). Haematological and physiological responses to sub lethal concentration of cadmium in freshwater teleost, *Tilapia zillii*.Wat. Air. Soil. Poll. 64(3): 551-558

Ghazala, M. S.; Al-Ghanim, K. A.; Sultana, S.; Al-Balawi, H. F. A.; Sultana, T.; Al-Misned, F.; and Ahmed, Z. (2014). Acute toxicity II: effect of organophosphates and carbamates to *Catla catla* fingerlings. JAPS, 24: 1785-1801.

Ghosh, M.; Manivannan, J.; Sinha, S.; Chakraborty, A.; and Mallick, S. K. (2012). In vitro and in vivo genotoxicty of silver nanoparticles. Mut. Res. Gen. Toxicol. Environ. Mutagen., 748 (1- 2): 60-68.

Gupta, P. K. (1977). Haematological techniques. 4th ed. Syndicate, pp.231.

Hamed, H. S.; and Osman, A. G. M. (2017). Modulatory effect of lycopene against carbofuran toxicity in African catfish (*Clarias gariepinus*). Fish Physiol. Biochem., 43:1721-1731.

Handy, R.D.; Henry, T.B.; Scown, T.M.; Johnston, B.D.; and Tyler, C.R. (2008). Manufactured nanoparticles: their uptake and effects on fish—a mechanistic analysis. Ecotoxicology 17(5), 396–409.

Hao, L.; and Chen, L. (2012). Oxidative stress responses in different organs of carp (*Cyprinus carpio*) with exposure to ZnO nanoparticles. Ecotoxicol. Environ. Saf., 80:108-110.

Hao, L.; Chen, L.; Hao, J.; and Zhong, N. (2013). Bioaccumulation and sub-acute toxicity of zinc oxide nanoparticles in juvenile carp (*Cyprinus carpio*): A comparative study with its bulk counterparts . Ecotoxicol. Environ. Saf.(22) 556-563

Harsij, M.; Paknejad, H.; Khalili, M.; Jafarian H.; and Nazari, S. (2021). Histological study and evaluation of Hsp70 gene expression in gill and liver tissue s of goldfish (*Carassius auratus*) exposed to Zinc oxide nanoparticles. Iran. J. Fish. Sci. 20: 741 - 760

Hassan, I.I.; Askar, A.A.; and El-Shourbagy, G.A.; (2004). Influence of some medicinal plants on performance; physiological and meat quality traits of broiler chicks. Egypt Poult. Sci. 24: 247–266.

Hawkin, R. I.; and Mawdesley, T. L. (2006). Fish aematology- a bibliography. J. Fish. Biol., 4: 183- 232.

Hedayati, A.; Kolangi, H.; Jahanbakhshi, A.; and Shaluei, F. (2012). Evaluation of silver nanoparticles ecotoxicity in silver carp (*Hypophthalmicthys molitrix*) and goldfish (*Carassius auratus*). Bulga. J. Vet. Med., 15 (3): 172-177.

Hedayati, S. A.; Farsani, H. G.; Naserabad, S. S.; Hoseinifar, A. H.; and Doan, H. V. (2018). Protective effect dietary vitamin E on immunological and biochemical induction through silver nanoparticles (Ag-NPs) inclusion in diet and silver salt ($AgNO_3$) exposure on zebrafish (*Dnio rerio*) Comp. Biochem. Physiol. part C: Toxicol. Pharmacol., 222: 100-107.

Huda, M.N.; Lewis, Z.; Kalanetra, K.M.; Rashid, M.; Ahmad, S.M.; Raqib, R.; Qadri, F.; Underwood, M.A.; Mills, D.A.; and Stephensen, C.B. (2014). Stool microbiota and vaccine responses of infants. Pediatrics.134(2):e362-72.

Hussain, S.; Boland, S. and Baeza-Squiban, A. (2008). Oxidative stress and proinflammatory effects of carbon black and titanium dioxide nanoparticles: role of particle surface area and internalized amount. J. Toxicol. 260:142–148.

Ibrahim, K. F. L.; AL-kelany, Y. L.; El-barasi, O. M.; Gheth, E. M. M.; and Eldurssi, I. S. (2021). Effect of zinc oxide nanoparticles on liver of male mice: Histological studies. Int. j. multidiscip. sci. adv. technol., 1: 346–354.

Ismail, S.H.; Hamdy, A.; Ismail, T.A.; Mahboub, H.H.; Mahmoud, W.H.; and Daoush, W.M. (2021). Synthesis and characterization of antibacterial carbopol/ZnO hybrid nanoparticles gel. Crystals 11, 1082.

Ismail, S.H.; Mohamed, G.; Amer, A.; and Amer, M. (2020). Comparative killing activity of different nanoparticles and nanocomposites based on Dermanys susgallinae. Nano Biomed. Eng. 12 (4): 338–350.

Jacobsen, N. R.; Stoeger, T.; Brule, S. V. D.; Saber, A. T.; Beyerle, A.; Vietti, G.; Mortensen, A.; Szarek, J.; Budtz, H. C.; Kermanizadeh, A.; Banerjee, A.; Ercal, N.; Vogel, U.; Wallin, H.; and Møller, P. (2015). Acute and subacute pulmonary toxicity and mortality in mice after intratracheal instillation of ZnO nanoparticles in three laboratories. Food Chem. Toxicol. 85: 84-85.

James, S.A.; Feltis, B.N.; deJonge, M.D.; Sridhar, M.; Kimpton, J.A.; Altissimo, M.; Mayo, S.; Zheng, C.;Hastings, A.; Howard, D.L.; Paterson, D.J.; Wright, P.F.A.; Moorhead, G.F.; Turney, T.W.; and Fu, J. (2013). Quantification of ZnO nanoparticle uptake, distribution, and dissolution within individual human macrophages. ACS Nano 7: 10621–35.

Jayaseelan, C.; Abdul Rahuman, R.; Ramkumar R.; Pajakumar, G.; Santhoshkumar, T.; and Marimuthu, S. (2014). Effect of sub-acute exposure to nickel nanoparticles on oxidative stress and histopathological changes in Mozambique tilapia, *Oreochromis mossambicus.* Ecotoxicol. Environ. Saf., 107: 220–228.

Jin, Y. X.; Zhang, X. X.; Shu, L. J.; Chen, L. F.; Sun, L.W.; Qian, H.; Liu, W.; and Fu, Z. (2010). Oxidative stress response and gene expression with atrazine exposure in adult female zebra fish (*Danio rerio*). Chemosphere, 78: 846-852.

Johari, S. A.; Kalbassi, M. R.; Soltani, M.; and Yu, I. J. (2013). Toxicity comparison of colloidal silver nanoparticles in various life stages of rainbow trout (*Oncorhynchus mykiss*). Iran. J. Fish. Sci. 12(1), 76 -85.

Kadry, S. M.; Marzouk, M. S.; Amer, A.F.; Hanna, M. I.; Azmy, A. H.; and Hamed, H. S. (2012). Vitamin E as antioxidant in female African catfish (*Clarias gariepinus*) exposed to chronic toxicity of atrazine. Egypt. J. Aquat. Biol. Fish., 16: 83-88.

Kalnie, S.J.; Alvarez, P.J.J.; Batley, G.E.; Fernandes, T.F.; Handy, R.D.; Lyon, D.Y.; Mahendra, S.; McLaughlin, M. J.; and Lead, J.R. (2006). Nanoma- terials in the environment: behavior, fate, bioavailability, and effects. Environ. Toxicol. Chem. 27: 1825-1851.

Kanaev, A.F. (1985). Veterinary Hygiene in Fish Farming. Moscow., 140-184.

Kargin, D. (2021). Changes in serum physiological and biochemical parameters of male swiss albino mice after oral administration of metal oxide nanoparticles (ZnO, CuO, and ZnO+CuO). Biol. Trace Elem. Res., 188:4218–4224

Kavitha, P. and Rao, J. V. (2007). Oxidative and locomotor behavior response as biomarkers for assessing recovery status of mosquito fish, *Gambusia affinis,* after lethal effect of an organo-phosphate pesticide, monocrotophos. Pest. Biochem. Physiol., 87: 182-188.

Kaya, H.; Aydın, F.; Gürkan, M.; Yılmaz, S.; Ates, M.; Demir, V.; and Arslan, Z. (2015). Effects of zinc oxide nanoparticles on bioaccumulation and oxidative stress in different organs of tilapia, *Oreochromis niloticus*. Environ. Toxicol. Pharmacol., 40(3): 836-847.

Kaya, H.; Aydın, F.; Gürkan, M.; Yılmaz, S.; Ates, M.; Demir, V.; and Arslan, Z. (2016). A comparative toxicity study between small and large size zinc oxide nanoparticles in tilapia, *Oreochromis niloticus*, Organ pathologies, osmoregulatory responses and immunological parameters. Chemosphere, 144: 571-582

Kaya, H.; Çelik, E.Ş.; Gürkan, M.; Yılmaz, S.; and Akbulut, M.; (2013). Effects of subchronic exposure to phosalone on oxidative stress and histopathological alterations in common carp (*Cyprinus carpio*, L.) J. Toxicol. Env. Heal. A., 76: 853-864.

Khalil, A. F.; Elkatry, H. O.; and El Mehairy, H. F. (2015). Protective effect of peppermint and parsley leaves oils against hepatotoxicity on experimental rats. Ann. Agric. Sci. 60: 353-358.

Khan, G. B.; Akhtar, N.; Khan, M. F. Ullah, Z.; Tabassum, S. and Tedesse, Z. (2022). Toxicological impact of Zinc Nano Particles on tilapia fish, *Oreochromis mossambicus*. Saudi J. Biol. Sci. 28: 1221-1226.

Khayal, E. E.; Ibrahim, H. M.; Shalaby, A. M.; Alabiad, M. A.; and El-Sheikh, A. A. (2021). Combined lead and zinc oxide-nanoparticles induced thyroid toxicity through 8-OHdG oxidative stress-mediated inflammation, apoptosis, and Nrf2 activation in rats. Environ. Toxicol., 36: 2588-2604.

Khodabakhshi, L.; Porbagher, H.; and Hosseini, S.V. (2017). The effect of time and concentration of nanoparticles of zinc oxide on histological pathology of rainbow trout gill. "J. Fish (IJNR)., 70(1), 84-85.

Kolarovic, J.; Popovic, M.; Zlinska, J.; Trivic, S.; and Vojnovic, M. (2010). Antioxidant activities of celery and parsley juices in rats treated with doxorubicin, Molecules, 15: 6183–6204.

Koner, D.; Banerjee, B.; Hasan, R.; and Saha, N. (2018). Antioxidant activity of endogenously produced nitric oxide against the zinc oxide nanoparticle-induced oxidative stress in primary hepatocytes of airbreathing catfish, *Clarias magur.* Nitric Oxide. 84: 7-15

Koohi, M. K.; Hejazy, M.; Najafi, D.; and Sajadi, S. M. (2017). Investigation of hematotoxic effect of nano ZnO, nano Fe_3O_4 and nano SiO_2 in vitro. Nanomed. Res. J., 2(2): 83-88.

Koracevic, D.; Koracevic, G.; Djordjevic, V.; Andrejevic, S.; and Cosic, V. (2001). Method for the measurement of antioxidant activity in human fluids. J. Clin. Pathol., 54: 356-361.

Kovacik, A.; Tvrda, E.; and Miskeje, M. (2018). Trace metal in the freshwater fish *Cyprinus carpio*: Effect to serum biochemistry and oxidative status markers. Biol. Trace. Elem. Res., 188: 484- 507.

Kumar, J.; Mitra, M. D.; Hussain, A.; and Kaul, G. (2018). Exploration of immunomodulatory and protective effect of *Withania somnifera* on trace metal oxide (zinc oxide nanoparticles) induced toxicity in *Balb/c* mice. Mol. Biol. Rep., 46:2447–2458

Laflamme, J.S.; Couillard, Y.; Campbell, P.G.; and Hontela, A.; (2000). Interrenal metallothionein and cortisol secretion in relation to Cd, Cu, and Zn exposure in yellow perch, *Perca flavescens*, from Abitibi lakes. Can. J. Fish. Aquat. Sci. 57: 1682–1700.

Lee, B.; Duong, C. N.; Cho, J.; Lee, J.; Kim, K.; Seo, Y.; Kim, P.; Choi, K.; and Yoon, J. (2012). Toxicity of citrate-capped silver nanoparticles in common carp (*Cyprinus carpio*). J. Biomed. Biotechnol. Article. 262670: 1–14.

Lee, J.; Kim, J.; Shin, Y.; Ryu, J.; Eom, I.; Lee, J.S.; Kim, Y.; Kim, P.; Choi, K.; and Lee, B. (2014) Serum and ultrastructure responses of common carp (*Cyprinus carpio L.*) during long-term exposure to zinc oxide nanoparticles. Ecotoxicol. Environ. Saf. 104: 8-17

Lefsrud, M.; Kopsell, D.; Wenzel, A.; and Sheehan, J. (2007). Changes in kale (*Brassica oleracea* L. var. *acephala*) carotenoid and chlorophyll pigment concentrations during leaf ontogeny. Sci. Hortic. 112: 136–141.

Li, H.C.; Zhou, Q.F.; Wu, Y.; Fu, J.J.; Wang, T.; and Jiang, G.B. (2008). Effects of waterborne nano-iron on medaka (*Oryzias latipes*): antioxidant enzymatic activity, lipid peroxidation and hispathology. Ecotoxicol. Environ. Saf. 72: 684–682.

Li, W.; Yin, D.; Zhou, Y.; Hu, S.; and Wang, L. (2003). 3, 4-Dichloroaniline- induced oxidative stress in liver of crucian carp (*Carassius auratus*). Ecotoxicol. Environ. Saf. 56:251-255.

Li, Y.J.; Song, T.B.; Cai, Y.Y.; Zhou, J.S.; Song, X.; Zhao X.; and Wu, X.L. (2008). Bisphenol A exposure induces apoptosis and upregulation of Fas/FasL and caspase-3 expression in the testes of mice. Toxicol. Sci.108(2):427-436.

Lopes, S.; Ribeiro, F.; Wojnarowicz, J.; Łojkowski, W.; Jurkschat, K.; Crossle, A.; Soares, A.M.V.M.; and Loureiro, S. (2013) Zinc oxide nanoparticles toxicity to Daphnia magna: size-dependent effects and dissolution Environ. Toxicol. Chem. 33: 180-188

Lozano-Fernandez, T.; Ballester-Antxordoki, L.; Perez-Temprano, N.; Rojas, E.; Sanz. D.; Iglesias-Gaspar, M.; Moya, S.; González-Fernández, Á.; and Rey, M. (2014). Potential impact of metal oxide nanoparticles on the immune system: the role of integrins, L-selectin and the chemokine receptor CXCR4. Nanomed. 10:1301–10.

Luo, M.; Shen, C.; Feltis, B.N.; Martin, L.L.; Hughes, A.E.; Wright, P.F.; and Turney, T.W. (2014). Reducing ZnO nanoparticle cytotoxicity by surface modification. Nanoscale 6:5781–5788.

Luoma, S.N.; and Rainbow, P.S. (2008). Sources and cycles of trace metals. In: Metal Contamination in Aquatic Environments: Science and Lateral Management. Cambridge University Press, Cambridge, pp. 47–66.

Ma, H.; Williams, P. L.; and Diamond, S.A. (2013) Ecotoxicity of manufactured ZnO nanoparticles–a review. Environ. Pollut. 172: 76–85.

Macleod, A. J.; Snyder, C. H.; and Subramanian, G. (1985) Volatile aroma constituents of parsley leaves, Phytochemistry, 24(11): 2623–7.

Mahboub, H. H.; Rashidian, G.; Hoseinifar, S. H.; Elsheshtawy, H. M.; Fahim, A.; Farsani, H. G.; Algharib, S. A.; and Doan, H. V. (2022). Protective effects against foodborne toxicity of zinc oxide nanoparticles in common carp (*Cyprinus Carpio*). Comp. Biochem. Physiol. C Toxicol. Pharmacol., 257:109345.

Maheswaran, R.; Devapanl, A.; Muralidharan, S.; Velmurugan, B.; and Ignaeimuthu, S. (2008). Haematological studies of fresh water fish, *Clarias batradrus (L)* exposed to mercuric chloride. Int. J. of Integr. Biol., 2(1): 48-54.

Mahmoud, M. A. M.; Yahia, D.; Abdel-Magiud, D. S.; Darwish, M. H. A.; Abd-Elkareem, M.; and Mahmoud, U. T. (2021). Broiler welfare is preserved by long-term low-dose oral exposure to zinc oxide nanoparticles: preliminary study. Nanotoxicology., 15(5): 605.

Maita, M.; Shiomitsu, K.; and Ikeda, Y. (1984). Health assessment by the climogram of hemochemical constituents in cultured yellow tail. Bull. Japn. Soc. Sci. Fish., 51: 205-211.

Manjula, D.; Dana, A. T.; John, T. B.; and Milton, F. (1986). Accumulation, physiological and biochemical effect of cadmium in a simple aquatic food chain. Ecotoxicol. Environ. Saf., 33 (1): 38-43.

Marisa, I.; Matozzo, V.; Munari, M.; Binelli, A.; Parolini, M.; Martucci, A.; Franceschinis, E.; Brianese, N.; and Marin, M. G. (2016). In vivo exposure of the marine clam *Ruditapes philippinarum* to zinc oxide nanoparticles: responses in gills, digestive gland and haemolymph. ESPR., 23: 15275–15283.

Masciangioli, T.; and Zhang, W.X. (2003). Environmental technologies at the nanoscale, Environ. Sci. Tochnol., 37: 102–108.

Maxine, M.; and Benjamine, B. S. (1985). Outline of veterinary clinical pathology 3 rd ed. Colorado State Univ. USA.

Mayer, A.; Vadon, M.; Rinner, B.; Novak, A.; Wintersteiger, R.; and Fröhlich, E. (2008). The role of nanoparticle size in hemocompatibility. Toxicology., 258, 138–147.

McLoughlin, N.; Yin, D.; Maltby, L.; Wood, R.M.; and Yu, H.; (2000). Evaluation of sensitivity and specificity of two crustacean biochemical biomarkers. Environ. Toxicol. Chem., 18: 2085–2082.

Moin, V.M. (1986): A simple and specific method for determining glutathione peroxidase activity in erythrocytes, Lab. Delo., 12: 724 -727

Mondal, A.H.; Behera, T.; Swain, P.; Das, R.; Sahoo, S.N.; Mishra, S.S.; Das, J.; and Ghosh, K. (2020). Nano zinc vis-à-vis inorganic Zinc as feed additives: Effects on growth, activity of hepatic enzymes and non-specific immunity in rohu, Labeo rohita (Hamilton) fingerlings. Aquac. Nutr., 26: 1211-1222.

Montgomery, D. W.; Simpson, S. D.; Engelhard, G. H.; Birchenough, S. N. R.; and Wilson, R. W. (2018). Rising CO_2 enhances hypoxia tolerance in a marine fish. Sci. Rep., 8:15152

Moreno, I.; Pichardo, S.; Jos, A.; Gómez-Amores, L.; Mate, A.; Vazquez, C. M.; and Cameán, A. M. (2005). Antioxidant enzyme activity and lipid peroxidation in liver and kidney of rats exposed to microcystin-LR administered intraperitoneally. Toxicon., 45:385–402

Moss, D.W. (1997) Physiochemical and pathophysiological factors in the release of membrane-bound alkaline phosphatase from cells., Clin. Chim. Acta., 257 (1): 133-40.

Munir, T.; Latif, M.; Mahmood, A.; Malik, A.; and Shafiq, F.(2020) Influence of IP-injected ZnO-nanoparticles in Catla catla fish: hematological and serological profile. Naunyn Schmiedebergs Arch. Pharmacol., 383: 2453–2461

Naguib, M.; Mohamad, U. M.; Mekkawy, I. A.; and Sayed, A.E.D. (2020). Hepatotoxic effect of silver nanoparticles on *Claries gariepinus*; Biochemical, histopathological, and histochemical studies. Toxicol. Rep., 7: 133-141.

Nieves-puigdoller, K.; Bj-rnsson, B.T.; and Mc-Cormick, S. D. (2007). Effect of hexazinone and atrazine on the physiology and endocrinology of smolt development in Atlantic salmon. Aquat. toxicol., 84: 27-37.

Nishikimi, M.; Appaji, N.; and Yagi, K. (1972). The occurrence of superoxide anion in the reaction of reduced phenazine methosulfate and molecular oxygen. Biochem. Biophys. Res. Commun., 46: 848-854.

Nohynek, G.J.; Lademann, J.; Ribaud, C.; and Roberts, M.S. (2007). Grey Goo on the skin? Nanotechnology, cosmetic and sunscreen safety, Crit. Rev. Toxicol., 37: 251–277.

Nusetti, O.; Esclapes, M.; Salazar, G.; Nusetti, S.; and Pulido, S. (2001). Biomarkers of oxidative stress in the polychaete *Eurthoecom planata* (Amphinomidae) under short-term copper exposure. Bull. Environ. Contam. Toxicol., 66: 76–81.

Ohkawa, H.; Ohishi, N.; and Tagi, K. (1978). Assay for lipid peroxides in animal tissues by thiobarbituric acid reaction. Ana. Biochem., 85: 351–358.

Önenç, S.S.; and Açıkgöz, Z. (2005). Aromatik bitkilerin hayvansal ürünlerde antioksidan etkileri. Hayvansal Üretim, 46: 50-55.

Orbea, A.; Fahimi, H. D.; and Cajaraville, M. P. (2000). Immunolocalization of four antioxidant enzymes in digestive glands of mollusks and crustaceans and fish liver. Histochem. Cell. Biol., 114: 383–404

Orsonneau, J. L.; Douet, P.; Massoube, C.; Lustenberger, P. and Bernard, S. (1989). An improved pyrogallol red-molybdate method for determining total urinary protein. Clin. Chem., 35: 2233-2236.

Pandurangan, M.; and Kim.D. (2015) ZnO nanoparticles augment ALT, AST, ALP and LDH expressions in C2C12 cells. Saudi J. Biol. Sci., 22: 678-684.

Park, J.; Lee, E.; Park, S.; Seong, N.; Kim, J., Kim, G.; Meang, E.; Hong, J.; Kim, S.; Koh, S.; Kim, M.; Kim, C.; Kim, S.; Son, S.; Seo, Y.; Kang, B.; Han, B.; An, S.; Yun, H.; and Kim, M. (2014). Toxicity of 100 nm zinc oxide nanoparticles: a report of 80-day repeated oral administration in Sprague Dawley rats. Int. J. Nanomed., 8: 108–126

Phull, A.R.; Nasir, B.; Haq, I. U.; and Kim, S. J. (2018). Oxidative stress, consequences and ROS mediated cellular signaling in rheumatoid arthritis. Chem. Biol. Interact., 281:121-36.

Piccinno, F.; Gottschalk, F.; Seeger, S.; and Nowack, B. (2012). Industrial production quantities and uses of ten engineered nanomaterials for Europe and the world. J. Nanopart. Res., 14:1-11.

Pokorny′, J.; and Pánek, J. (2012). The effect of natural antioxidants in herbs and spices on food shelf-life., Handbook of Herbs and Spices (2nd ed). 2: 51–71.

Popovic, M.; Kaurinovic, B.; Jakovljevic, V.; Mimica-Dukic, N.; and Bursac, M. (2007). Effect of parsley (*Petroselinum crispum* (Mill.) Nym. ex A.W. Hill, Apiaceae) extracts on some biochemical parameters of oxidative stress in mice treated with CCl_4, Phytother. Res., 21: 717–723.

Prenter, J.; MacNeila, C.; Dicka, J.T.A.; Riddella, G.E.; and Dunn, A.M. (2004). Lethal and sublethal toxicity of ammonia to native, invasive and parasitized freshwater amphipods. Water. Res., 38: 2847-2850

Prentera, J.; MacNeila, C.; Dicka, J.T.A.; Riddella, G.E.; and Dunnb, A.M. (2004) . Lethal and sublethal toxicity of ammonia to native, invasive and parasitized freshwater amphipods.Water Res., 38(12): 2847-2850

Rahdar, A.; Hajinezhad, M.; Sivasankarapillai,V.; Askarim, F.; Noura, M.; and Kyzas, G. (2020) . Synthesis, characterization, and intraperitoneal biochemical studies of zinc oxide nanoparticles in *Rattus norvegicus*. Appl. Phys. A. 126:347

Rahmat, A.; Ahmad, N.S.S.; and Ramli, N.S. (2018). Parsley (*Petroselinum crispum*) supplementation attenuates serum uric acid level and improves liver and kidney structures in oxonate-induced hyperuricemic rats. Orient. Pharm. Exp. Med. 18**:** 383–401.

Rajaraman, V.; Nonnecke, B.J.; Franklin, S.t.; Hamell, D.C.; and Horst, R.L. (1998): Effect of vitamin A and E on nitric oxide production by mononuclear leukocytes from neonatal calves fed milk replacer. J. Dairy Sci., 81:3278-3285

Rajkumar, K. S.; Sivagaami, P.; Ramkumar, A.; Murugadas, A.; Srinivasan, V.; Arun, S.; Kumar, P. S.; and Thirumurugan, R. (2022). Bio-functionalized zinc oxide nanoparticles: Potential toxicity impact on freshwater fish (*Cyprinus carpio)*. Chemosphere., 280: 133220

Rajkumar, K.S.; Kanipandian, N.; Thirumurugan, R. (2015). Toxicity assessment on haemo-tology, biochemical and histopathological alterations of silver nanoparticles exposed fresh water fish, *Labeo rohita*, Appl. Nanosci., 6: 18–28.

Rani, V.; Verma, Y.; and Rana, S. V. S. (2021). Zinc oxide nanoparticles ameliorate dimethylnitrosamine-induced renal Toxicity in rat. Appl. Biochem. Biotechnol., 4:1699-1715.

Rashidian, G.; Lazado, C. C.; Mahboub, H. H.; Mohammadi-Aloucheh, R.; Proki´, M. D.; Nada, H. S.; and Faggio, C. (2021). Chemically and green synthesized ZnO nanoparticles alter key immunological molecules in common carp, *Cyprinus carpio*, Skin Mucus. Int. J. Mol. Sci., 22: 3270

Rathore, R. S.; and Khangarot, B. S. (2002). Effect of temperature on the sensitivity of sludge worm Tubifex tubifex (Muller) to selected heavy metals. Ecotoxicol. Environ. Saf., 53: 27-36.

Reddy, P. K.; and Leatherland, J. F. (1988). Stress physiology, in: fish disease and disorders, Vol.2 Non-infection Disorders. (Eds. Leatherland, J. F. & Woo, P. T. K.). Cabi. Pub., 278-301.

Reza, E. H.; Kobra, S.; Leila, S.; Vahid, Y.; and Esmaiel, A. (2014). Investigation of the zinc oxide nanoparticles effect on testosterone, cholesterol and cortisol in rats. Res. J. Recent Sci. 3(4), 14-18.

Roshanai, K.; Razavian, M.; Ahmadi, R.; Heidarieh, N.; and Masaeemanesh, M. (2012). The effect of silvernano oral consumption on some hormonal, hematological and urine parameters of vistar rats. Qom. Univ.Med. Sci. J. 6 (3) :65-70.

Ruiz, A.; Morais, P.C.; De-Azevedo, R.B.; Lacava, Z.G.; Villanueva, A.; and Morales M. P. (2014). Magnetic nanoparticles coated with dimercaptosuccinic acid: development, characterization, and application in biomedicine. J. Nanopart. Res., 16(11):2588.

Saddick, S.; Afifi, M.; and Abu Zinada, O. A. (2015). Effect of Zinc nanoparticles on oxidative stress-related genes and antioxidant enzymes activity in the brain of *Oreochromis niloticus* and *Tilapia zillii.* Saudi J. Biol. Sci. 24: 1672-1678.

Sakai, T.; Murata, H.; Yamauchi, K.; Sekiya, T.; and Ukawa, M. (1992). Effects dietry lipid peroxides contents on *in vitro* lipid peroxidation, α - tochopherol contents and superoxide dismutase and glutathione peroxidase activities in the liver of yellow tilapia. Nippon Suisan Gakkaishi, 58: 1483-1486.

Saman, S.; Moradhaseli, S.; Shokouhian, A.; and Ghorbani, M. (2013). Histopathological effects of ZnO nanoparticles on liver and heart tissues in wistar rats. Adv. Biores., 2: 83- 88.

Samipillai, S.; and Jagadeesan, G. (2005). Influence of taurine on phosphatases activity in selected tissues of mercuric chloride intoxicated mice. J. Exp. Zool., 8: 285-300.

Scheletter, G.; and Nussel, E. (1975). Arbeitsmed sozialmed pracentimed, 28: 10-25.

Scott, G.; and Sloman, A. (2004) The effects of environmental pollutants on complex fish behaviour: integrating behavioural and physiological indicators of toxicity. Aquat. Toxicol., 68: 368–382.

Sener, G.; Saçan, Ö.; Yanardag, R.; and Ayanoglu-Dülger, G. (2003). Effects of parsley (*Petroselinum crispum*) on the aorta and heart of Stz induced diabetic rats, Plant Foods Hum. Nutr. 58 : 1–7.

Shahzad, K.; Khan, M.N.; Jabeen, F.; Kosour, N.; Chaudhry, A. S.; Sohail, M.; and Ahmad, N. (2018). Toxicity of zinc oxide nanoparticles (ZnO-NPs) in tilapia (*O. mossambicus*): tissue accumulation, oxidative stress, histopathology and genotoxicity. Int. J. Environ. Sci. Technol., 16:1873–1884.

Shaluei, F.; Hedayati, A.; Jahanbakhshi, A.; and Baghfalaki, M.; (2012). Physiological responses of great sturgeon (*Huso huso*) to different concentrations of 2-phenoxyethanol as an anesthetic. Fish. Physiol. Biochem. 38: 1627–1634.

Shaluei, F.; Hedayati, A.; Jahanbakhshi, A.; Kolangi, H.; and Fotovat, M. (2013). Effect of subacute exposure to silver nanoparticle on some hematological and plasma biochemical indices in silver carp (*Hypophthmichthys molitrix*). Hum. Exp. Toxicol., 32: 1270-1277.

Shen, C.; James, S.A.; de Jonge, M.D.; Turney, T.W.; Wright, P.F.; and Feltis, B.N.; (2013). Relating cytotoxicity, zinc ions, and reactive oxygen in ZnO nanoparticle-exposed human immune cells. Toxicol. Sci., 136:120–30.

Singh, N.; Manshian, B.; and Jenkins, G.J.S. (2008). Nano genotoxicology: the DNA damaging potential of engineered nanomaterials. Biomaterials 30(23):3881–3814.

SPSS (2018) SPSS for windows (version 26) Armonk, Illinois:SPSS. Inc

Srikanth, K.; Pereira, E.; Duarte, A. C.; Ahmad, I.; and Rao, J. V. (2015). Assessment of cytotoxicity and oxidative stress induced by titanium oxide nanoparticles on chinook salmon cells. ESPR., 22: 15571–15578.

Srivastav, A.; Kumar, M.; Ansari, N.; Jain, A.; Shankar, J.; Arjaria, N.; Jagdale, P.; and Singh, D. (2016). A comprehensive toxicity study of zinc oxide nanoparticles versus their bulk in Wistar rats: Toxicity study of zinc oxide nanoparticles. Hum. Exp. Toxicol .(35) 1286–1304.

Stein, E. A. (1986). In textbook of clinical chemistry, W. B. Saunder, Philadelphia, 28: 878-886.

Stoskopf, M.F. (1983). fish medicine. Sounders Company, U.S.A. Pp882.

Sudhakaran, S.; Athira, S.S.; Suresh, B.S.; Varma, H.K.; and Mohanan, P.V. (2020). Determination of the bioavailability of zinc oxide nanoparticles using ICP-AES and associated toxicity. Colloids Surf. B. 188:110767.

Suganthi, P.; Murali, M.; Athif, P.; Sadiq Bukhari, A.; Syed Mohamed, H.E.; Basu, H.; Singhal, R. K. (2018). Haemato-immunological studies in ZnO and TiO_2 nanoparticles exposed euryhaline fish, *Oreochromis mossambicus*. Environ. Toxicol. Pharmacol., 66: 55-61.

Suganthi, P.; Murali, M.; Sadiq Bukhari, A.; Basu, H.; and Singhal, R. K. (2015). Haematological studies on freshwater tilapia treated with ZnO nanoparticles. J. adv. App. Sci. Res., 1: 41–67.

Suman, T.Y.; Radhika Rajasree, S.R.; and Kirubagaran, R. (2015). Evaluation of zinc oxide nanoparticles toxicity on marine algae *chlorella vulgaris* through flow cytometric, cytotoxicity and oxidative stress analysis. Ecotoxicol. Environ. Saf., 113, 23–30.

Suvarna, S. K.; Layton, C.; and Bancroft, J. D. (2018). Bancroft's Theory and Practice of Histological Techniques. Eighth edition. Oxford: Elsevier.

Syama, S.; Reshma, S. C.; Sreekanth, P. J.; Varma, H. K.; and Mohanan, P. V. (2013). Effect of zinc oxide nanoparticles on cellular oxidative stress and antioxidant defense mechanisms in mouse liver. Toxicol. Environ. Chem., 85: 485- 503.

Syuichi, S.; Yone, Y. (1978) Effect of starvation on hematological characteristics, and the contents of chemical components and activities of enzymes in blood serum of red Sea Bream . J. Fac. Agric. Kyushu Univ., 23: 63-68.

Taheri, S.; Banaee, M.; Haghi, B. and Mohiseni, M. (2017). Effects dietary supplementation of zinc oxide nanoparticles on some biochemical biomarkers in common carp (*Cyprinus carpio*). Int. J. Aquat. Biol., 5: 286-284.

Taherian, S.M.R.; Hosseini, S.A.; Jafari, A.; Etminan, A.; and Birjandi, M. (2018) Acute toxicity of zinc oxide nanoparticles from satureja hortensis on rainbow trout ,*Oncorhynchus mykiss*, Turk. J. Fish.& Aquat. Sci., 20 (6): 481-488.

Thwala, M.; Musee, N.; Sikhwivhilu, L.; and Wepener, V. (2013). The oxidative toxicity of Ag and ZnO nanoparticles towards the aquatic plant *Spirodela punctuta* and the role of testing media parameters. Environ. Sci.: Processes Impacts., 15: 1830-1843.

Tietz, N. W. (1984). Textbook of Clinical Chemistry, 2 nd ed. W. B. Saundes Co., Philadelphia., 831-832.

Tietz, N. W. (1986). Textbook of Clinical Chemistry, W. B. Saunders Company, Philadelphia., 28: 1270- 1271.

Tietz, N.M.; and Norbert, W. (1985). Clinical Guide to Laboratory Tests, 3 rd Edition, WB Saunders, Philadelphia., 186-188.

Vajargah, M. F.; Imanpoor, M. R.; Shabani, A. and Hedayati, A. (2018). Effect of long-term exposure of silver nanoparticles on growth indices, hematological and biochemical parameters and gonad histology of male goldfish (*Crassius auratus gibelio*). Microsc. Res. Tech., 82 (7): 1224-1230.

Valavanidis, A.; Vlahogianni, T.; Dassenakis, M.; and Scoullos, M. (2006). Molecular biomarkers of oxidative stress in aquatic organisms in relation to toxic environmental pollutants. Ecotoxicol. Environ. Saf., 64: 178-188.

Van Der Merwe, M. (1992). Aspects of heavy metal Concentration in the Olifants River, Kruger National Park, and the effect of copper on the haematology of *Clarias gariepinus* (Clariidae). M.Sc. thesis, Rand Afrikaans University. South Africa.

Vandebriel, R. J.; Tonk, E. C. M.; Fonteyne-Blankestijn, L. J.; Gremmer, E. R.; Verharen, H. W.; van der Ven, L. T.; van Loveren, H.; and Jong, W. H. (2014) Immunotoxicity of silver nanoparticles in an intravenous 28-day repeated-dose toxicity study in rats. Part Fibre Toxicol. 11: 21–28

Velisek, J.; Stara, A.; Kolarova, J.; and Svobodova, Z. (2011). Biochemical, physiological and morphological responses in common carp, (*Cyprinus carpio* L). after long term exposure to terbutryn in real environmental concentration. Pest. Biochem. Physiol., 100: 305-313.

Wang, C.; Wang, X.M.; Xu, B.Q.; Zhao, J.C.; Mai, B.X.; Peng, P.A.; Sheng, G.Y.; and Fu, J.M. (2004). Enhanced photocatalytic performance of nano sized coupled ZnO/SnO_2 photocatalysts for methyl orange degradation, J. Photoch. Photobio, (A)., 168: 47-52.

Wang, D.; Li, H.; Liu, Z.; Zhou, J.; and Zhang, T. (2017). Acute toxicological effects of zinc oxide nanoparticles in mice after intratracheal instillation. Int. J. Occup. Environ. Health., 23:11-18.

Wang, L.; Ding, W.; and Zhang, F. (2012). Acute toxicity of ferric oxide and zinc oxide nanoparticles in rats . J. Nanosci. Nanotechnol., 10: 8617-8624.

Wigginton, N.S.; Haus, K.L.; and Hochella, M.F. (2007) Aquatic environmental nanoparticles, J. Environ. Monit., (8). 1306-1316.

Wong, P.; and kitts, D. (2006). Studies on the dual antioxidant and antibacterial properties of parsley (*Petroselinum crispum*) and cilantro (*Coriandrum sativum*) extracts. Food Chem. 87: 505- 515.

Wu, Y.; and Zhou, Q. (2013). Silver nanoparticles cause oxidative damage and histological changes in medaka (*Oryzias latipes*) after 14 days of exposure. Environ. Toxicol. Chem., 32(1) :165-173.

Xia, T.; Kovochich, M.; Brant, J.; Hotze, M.; Sempf, J.; Oberley, T.; Sioutas, C.; Yeh J.I.; Wiesner, M.R.; and Nel, A.E.; (2006). Comparison of the abilities of ambient and manufactured nanoparticles to induce cellular toxicity according to an oxidative stress paradigm. Nano let., 6(8): 1784-1807.

Xia, T.; Kovochich, M.; Liong, M.; Madler, L.; Gilbert, B.; Shi, H.; Yeh, J. I.; Zink, J. I. and Nel, A. E. (2008) Comparison of the mechanism of toxicity of zinc oxide and cerium oxide nanoparticles based on dissolution and oxidative stress properties. ACS Nano 2(10):2121–2134.

Xiong, D.; Fang, T.; Yu, L.; Sima, X. and Zhu, W. (2011). Effects of nano-scale TiO2, ZnO and their bulk counterparts on zebrafish: Acute toxicity, oxidative stress and oxidative damage. Sci. Total Environ., 409(8): 1444-1452

Yahya, R. A. M.; Attia, A. M.; El-Banna, S. G.; El-Trass, E. E.; Azab, A. E.; JM, J.; and El.M.Shkal, K. (2018). Hematotoxicity induced by copper oxide and/or zinc oxide nanoparticles in male albino rats. Biotechnol. Bioeng., 3:1-7.

Yan, G.; Huang, Y.; Bu, Q.; Lv, L.; Deng, P.; Zhou, J.; Wang, Y.; Yang, Y.; Liu, Q.; Cen, X.; and Zhao, Y. (2012). Zinc oxide nanoparticles cause nephrotoxicity and kidney metabolism alterations in rats. J. Environ. Sci. Health. A Tox Hazard. Subst. Enviro. Eng., 47(7): 577-588

Young, D. S. (1990). Effects of drugs on clinical laboratory tests, 3 rd Ed., 3: 168-182.

Yousafzai, A. M.; and Shakoori, A. R. (2011). Hepatic responses of a freshwater fish against aquatic pollution. Pakistan J. Zool., 43 (2): 208-221.

Yousef, M.; Mutar, T.; and Kamel, M. (2018). Hepato-renal toxicity of oral sub-chronic exposure to aluminum oxide and/or zinc oxide nanoparticles in rats. Toxicology Reports., 6: 336-346.

Yousefian, M.; and Payam, B. (2012). Effects of nanochemical particles on some histological parameters of fish. Adv. Environ. Bio., 6(3): 1208-1215.

Yu, X.; Zhao, X.; Ze, Y.; Wang, L.; Liu, D.; Hong, J.; and Li, B. (2014). Changes of serum parameters of TiO_2 nanoparticle-induced atherosclerosis in mice.J.Hazard. Mat., 280: 364-371.

Zhang, H.; Chen, F.; Wang, X.; and Yao, H.Y. (2006). Evaluation of antioxidant activity of parsley (*Petroselinum crispum*) essential oil and identification of its antioxidant constituents. Food Res. Int., 38: 833-838.

Zhao, X.; Wang, S.; Wu, Y.; You, H.; and Lv, L. (2013). Acute ZnO nanoparticles exposure induces developmental toxicity, oxidative stress and DNA damage in embryo-larval zebrafish. Aquat. Toxicol., 136: 48–58.

Zhu, X.; Wang, J.; Zhang, X.; Chang, Y.; and Chen, Y. (2008). The impact of ZnO nanoparticle aggregates on the embryonic development of zebrafish (*Danio rerio*). Nanotechnology., 20 (19):195103.

Zhu, X.; Zhu, L.; Duan, Z.; Qi, R.; Li, Y.; and Lang, Y. (2008). Comparative toxicity of several metal oxide nanoparticle aqueous suspensions to Zebrafish (*Danio rerio*) early developmental stage. J. Environ. Sci. Health (Part A) 43: 278–284.

Öner, M.; Atli, G.; and Canli, M. (2008). Changes in serum biochemical parameters of freshwater fish *Oreochromis niloticus* following prolonged metal (Ag, Cd, Cr, Cu, and Zn) exposure. Environ. Toxicol. Chemist., 27 (2): 360-366.

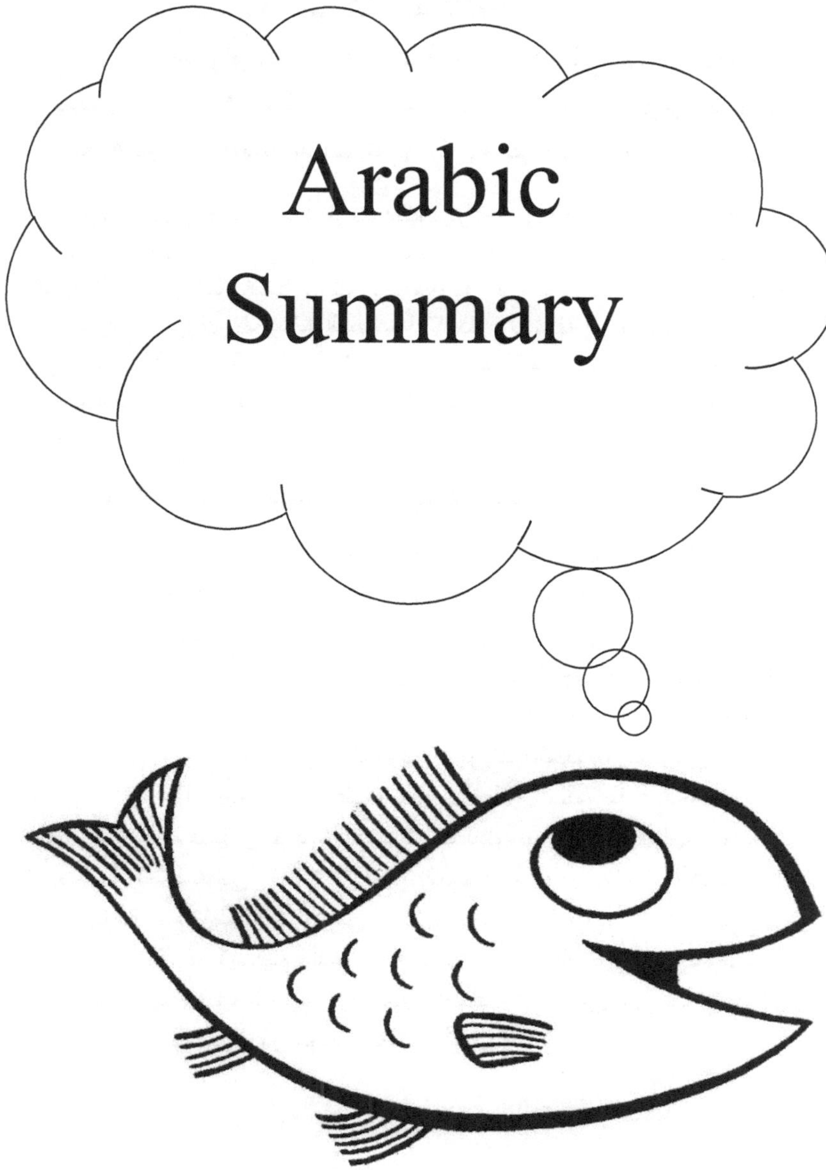
Arabic
Summary

الملخص العربي

دراسة فسيولوجية، هستولوجية، ومناعية على تأثير التغذية بمستخلص أوراق المقدونس و جزيئات المقدونس النانوميترية ضد سمية جزيئات أكسيد الزنك النانوميترية في أسماك البلطي النيلي

لتقييم تأثير جزيئات أكسيد الزنك النانوميترية علي القياسات الدموية والكيمياء الحيويه والأجهاد التاكسدي والأنزيمات المضاده للأكسده. أجريت هذه الدراسة أيضا لتقييم تأثير جزيئات اكسيد الزنك النانوميترية على قياسات الأجهاد التأكسدي الكيميائية والحيوية والانزيمات المضادة الأكسدة. بالأضافة إلي ذلك هدفت الدراسة إلي تصليت الضوء علي التغييرات النسيجية المحدثة من التعرض لجزيئات اكسيد الزنك النانوميترية . وإظهار دور اوراق المقدونس والنانو المقدونس في تحسين هذه التغيرات.

التصميم التجريبي والمعاملات التى تم فحصها:

في هذه الدراسة الحالية تم أستخدام ٤٠٠ سمكة من أسماك البلطي النيلي أخذت ٥٠ سمكة لتعيين التركيز نصف المميت لجزيئات أكسيد الزنك النانوميترية. تم تقسيم سمكة الى ستة مجموعات. كل مجموعة تتكون من ثلاث تكررات بحيث كل حوض يحتوي علي ١٠ أسماك . المجموعة الأولي كانت بمثابة المجموعة الضابطة. وعملت المجموعة الثانية وبينما مثلت المجموعة الثانية والثالثة كمجموعة الضابطة وتم إضافة ٤ جم من مستخلص أوراق المقدونس وجزيئات المقدونس النانوميترية / كيلو جرام من وزن الوجبه الغذائية علي التوالي. تعرض المجموعة الرابعة إلى١٠/١ التركيز النصف مميت من جزيئات أكسيد الزنك النانوميترية لمدة ٦ أسابيع . تعرضت المجموعة الخامسه إلى ١٠/١ التركيز النصف مميت من جزيئات أكسيد الزنك النانوميترية مع إضافة ٤ جم من مستخلص أوراق المقدونس / كيلو جرام من الوجبه الغذائية لمدة ٦أسابيع ، المجموعه السادسه ١٠/١ التركيز النصف مميت من جزيئات أكسيد الزنك النانوميترية مع إضافة ٤ جم من جزيئات المقدونس النانوميترية / كيلو جرام من الوجبه الغذائية.

بعد ٦ أسابيع من المعاملة، تم تجميع عينات الدم من الوريد. تم جمع مصل الدم واستئصلت أنسجة الكبد من أجل الدراسات البيوكيميائية والنسيجية.

تضمن العمل الحالي الفحوصات التالية:

-الفحص المرضي وما بعد الوفاه.

- المصل: تحديد (الفوسفاتيز القلوي، الالانين أمينو ترانسفيريزو أسبارتات أمينو ترانسفيراز) بالاضافه إلى وظائف الكلي (اليوريا &الكرياتينين) ،البروتين، ومستوى الجلوكوز فى الدم والكورتيزول

-ومستوى الدهون بالدم

- الفحص النسيجي للكبد

النتائج:

تم تعيين التركيز النصف المميت لجزيئات أكسيد الزنك النانوميترية حيث كان ٨ مجم/ لتر فى ٩٦ ساعة

العلامات الأكلينيكية ونتائج ما بعد الوفاه:

تتجلي العلامات الأكلينيكية لأسماك البلطي بعد تعرضه لسمية أكسيد الزنك (٨مجم/ لتر) بعد أظهر تصبغًا غير طبيعي للجلد ، والتحرك نحو مضخة الهواء ، والبقاء لفترة طويلة في مكان معين ، وحركة غير منتظمة ، وفقدان التوازن ، وفقدان الشهية أظهرت تراكم الدهون في الكبد مع شحوب في الأعضاء الداخلية الأخرى.

القياسات البيوكيميائية:

أظهرت التحاليل الكيميائية الحيوية زيادة فى قيم إنزيمات الكبد (الفوسفاتيز القلوي، الألانين أمينو ترانسفيريزو أسبارتات أمينو ترانسفيراز) بالاضافة إلى اليوريا، الكرياتينين ، البروتين، الألبيومين، الجلوبيولين، الجلوكوز، الكورتيزول، في المجموعة الرابعة المعرضة ل ٨ مجم/ لتر من جزيئات أكسيد الزنك النانوميترية لمدة ٦ اسابيع.

سجلت معاملات الأجهاد التأكسدي في أنسجة الكبد ما يلي:

أدى التعرض ل ٨ مجم/ لتر من جزيئات أكسيد الزنك النانوميترية في أسماك البلطي النيلي إلى إرتفاع في الأكسدة الفوقية للدهون السعة الكلية لمضادات الأكسدة و مالوندي الديهيد في أنسجة الكبد. فى حين أن في الكبد مستوى الجلوتاثيون والكتاليز و سوبرأكسيد ديزميوتيز وأكسيد النيتروز والجلوتاثيون بيروكسيديز انخفض ، مقارنة بالأسماك الضابطة .

تضمنت التغييرات النسيجية المرضية الأكثر ملاحظة ما يلي:

- **الكبد:** تراكم خلايا الكبد مع إلتهاب خلوي بالمسالك البابية ووجود تليف وإلتهاب بخلايا الكبد وبؤر متعدده.

الخلاصة:

حسنت التغذية بأوراق المقدونس والنانو المقدونس لأسماك البلطي النيلي من الآثار الجانبية لسمية أكسيد النانوميترية وتحسين المتغيرات السابقة بالقرب من المجموعة الضابطة. كشفت الدراسة الحالية أن النظام الغذائي بأوراق المقدونس والنانو المقدونس له تأثير جيد ضد سمية أكسيد الزنك النانوميترية في أسماك البلطي النيلي.

المستخلص العربى

تستخدم جزيئات أكسيد الزنك النانوميترية بشكل فعال في تطبيقات تربية الأحياء المائية المتنوعة .هدفت هذه الدراسة الى تعيين التركيز النصف مميت في أسماك سمك البلطي بالأضافة إلى الدور الوقائي لأوراق المقدونس والنانو المقدونس ضد السمية المحدثة من جزيئات أكسيد الزنك النانوميترية.

تم تقسيم الأسماك إلى ٦ مجموعات كل مجموعه ثلاث تكررات كالاتى: المجموعة الأولي وهى المجموعة الظابطة، المجموعة الثانية وعملت المجموعة الثانية كمجموعة ضابطة وتم إضافة ٤ جم من مستخلص أوراق المقدونس / كيلو جرام من وزن العليقة. تعرضت المجموعة الثالثة الى ٤ جم من جزيئات المقدونس النانوميترية / كيلو جرام من وزن الوجبة الغذائية. تم تعرض المجموعة الرابعة إلى ١٠/١ التركيز النصف مميت من جزيئات أكسيد الزنك النانوميترية لمدة ٦ أسابيع. تعرضت المجموعه الخامسه إلي ١٠/١التركيز النصف مميت من جزيئات أكسيد الزنك النانوميترية مع إضافة ٤ جم من مستخلص أوراق المقدونس / كيلو جرام من العليقة لمدة ٦أسابيع ، المجموعه السادسه ١/ ١٠ من التركيز النصف مميت من جزيئات أكسيد الزنك النانوميترية مع إضافة ٤ جم من جزيئات المقدونس النانوميترية / كيلو جرام من الوجبه الغذائيه لمدة٦ أسابيع. في نهاية التجربة تم جمع عينات الدم وأنسجة الكبد من الأسماك للتحاليل البيوكيميائية و الفحوصات النسيجية.

أظهرت النتائج زيادة ملحوظة في إنزيمات الكبد (الفوسفاتيز القلوي، الالنين أمينو ترانسفيريزو أسبارتات أمينو ترانسفيراز) ووظائف الكلي (اليوريا، الكرياتينين). سجلت النتائج إرتفاع في مستويات البروتينات الكلية، الالبيومين، الجلوبيولين، ونسبة الالبيومين / الجلوبيولين. كذلك في مستويات الجلوكوز و الكورتيزول ،و الإنزيم النازع هيدروجين بالدم. وتم تسجيل مستويات الدهون الكلية (الكوليسترول والدهون الثلاثية).

سجل مستوى الجلوتاثيون والكتاليز وسوبرأكسيد ديزميوتيز وأكسيد النيتروز والجلوتاثيون بيروكسيديز انخفاض ملحوظ ، زادت السعة الكلية لمضادات الأكسدة والاكسدة الفوقية للدهون. مقارنة بالأسماك في المجموعة الضابطة بعد التعرض ل ٨ ملجم / لتر من جزيئات أكسيد الزنك النانوميترية لمده ٦ اسابيع .

التحاليل الدموية:

اظهرت نقص في كرات الدم البيضاء ومتوسط تركيز الهيموجلوبين بكرات الدم الحمراء ومتوسط حجم كرات الحمراء والصفائح الدموية ، بينما زادت نسبه الهيموجلوبين و متوسط تركيز الهيموجلوبين بكرات الدم الحمراء.

التغيرات النسيجيه للكبد :

الدراسة الهستولوجية اوضحت تغيرات في أنسجة الكبد بأسماك البلطي المعرضة ل٨ ملجم/ لتر من جزيئات اكسيد الزنك النانوميترية لمدة ٦ اسابيع مقارنة بالمجموعة الضابطة والمجموعة التي تغذت علي أوراق المقدونس وجزيئات المقدونس النانوميترية (٤جم /كجم) من العليقة .

نستنتج من هذه الدراسة أن تغذية أسماك البلطي النيلي على ٤ جم من مستخلص أوراق المقدونس أو جزيئات المقدونس النانوميترية / كيلو جرام من العليقة قد حسنت من التأثيرات المدمرة لسمية جزيئات اكسيد الزنك وقربها من المجموعة الضابطة وكان ذلك أكثر وضوحاً في الأسماك التي تم معالجتها بجزيئات المقدونس النانوميترية .

الكلمات المفتاحية :

جزيئات أكسيد الزنك النانوميترية ، أسماك البلطي النيلي ، المقدونس ، جزيئات المقدونس النانومتيرية ، الكيمياء الحيوية ، التغيرات الباثولوجية.

FSC
www.fsc.org
MIX
Papier | Fördert
gute Waldnutzung
FSC® C083411

Kolibri 360 GmbH
Ferdinand-Jühlke-Straße 7
99095 Erfurt, Deutschland
produktsicherheit@kolibri360.de

Druck:
CPI Druckdienstleistungen GmbH
im Auftrag der
Kolibri 360 GmbH
Ferdinand-Jühlke-Str. 7
99095 Erfurt